INTERNET COMPANION FOR STATISTICS

GUIDE AND ACTIVITIES FOR THE WEB

The Companion Website can be accessed at
http://larsen.duxbury.com
Please enter the Serial Number from the
inside back cover when prompted

INTERNET COMPANION FOR STATISTICS

GUIDE AND ACTIVITIES FOR THE WEB

Michael D. Larsen
The University of Chicago

THOMSON
BROOKS/COLE

Australia • Canada • Mexico • Singapore • Spain • United Kingdom • United States

Printed in Canada
1 2 3 4 5 6 7 07 06 05 04 03

Printer: Transcontinental Printing, Inc. -Louiseville

ISBN: 0-534-42356-6

For more information about our products, contact us at:
Thomson Learning Academic Resource Center
1-800-423-0563

For permission to use material from this text, contact us by:
Phone: 1-800-730-2214
Fax: 1-800-731-2215
Web: http://www.thomsonrights.com

Cover Design: Yvo Riezebos

For more information contact:
Brooks/Cole-Thomson Learning
10 Davis Drive
Belmont, CA 94002-3098
USA

Asia
Thomson Learning
5 Shenton Way #01-01
UIC Building
Singapore 068808

Australia/ New Zealand
Thomson Learning
102 Dodds Street
South Street
Southbank, Victoria 3006
Australia

Canada
Nelson
1120 Birchmount Road
Toronto, Ontario M1K 5G4
Canada

Europe/Middle East/South Africa
Thomson Learning
High Holborn House
50/51 Bedford Row
London WC1R 4LR
United Kingdom

Latin America
Thomson Learning
Seneca, 53
Colonia Polanco
11560 Mexico D.F.
Mexico

Spain/ Portugal
Paraninfo
Calle/Magallanes, 25
28015 Madrid, Spain

Contents

Preface

Introduction and Purpose of the Book

The *Internet Companion for Statistics: Guide and Activities for the Web*, first edition, provides educators and students an organized, clear interface to material on the Internet useful for teaching and learning introductory statistics. The examples in the book are linked to motivating material on the Internet and organized according to specific course topics. Exercises include numerical, short answer, and expository problems related to the Internet sites. Some problems ask students to enter data, run an on-line application program, and comment on the results.

The Companion Website can be accessed at http://larsen.duxbury.com. Please enter the Serial Number from the back cover when prompted. The Companion Website presents the exercises in a convenient format for classroom and assignment use and includes solutions, additional links and problems, and a subject index.

Two major challenges to using the numerous resources on the Internet for teaching and learning is that material is seldom organized for a particular educational purpose and quality varies greatly across and within sites. The information on the Internet also rarely is accompanied by introductions related to statistical topics, explanations of advanced terminology, and study questions. This book provides the necessary organization and supplemental material to make it easy and enjoyable to incorporate the Internet into teaching and learning statistics.

If students are expected to be able to look at a newspaper article, a data set, or on-line material and report and justify a conclusion based on numerical and statistical information, then students should be asked to do this in assignments. The book includes many questions and writing assignments of this nature. The opportunity for exploration on the Internet and further reading on some Internet sites allows students and instructors to follow their interests and enhance their enjoyment of statistics.

Intended Audience

The primary intended audience is students and instructors of introductory statistics courses. Topics of such courses include describing data, elementary research design, probability, and statistical inference. Increasingly a non calculus version of introductory statistics is being taught in U.S. high schools. The number of students taking the College Board AP Statistics exam, which was first given in 1997, has grown dramatically. Colleges and universities offer courses in quantitative literacy and pre- and post-calculus introductory probability and statistics. This book, written for a general statistics and probability course, is appropriate for all of these audiences. Examples and sites are taken from many fields, including economics, engineering, environmental sciences, health and medicine, opinion polling, and social and political science. The wide applicability of basic ideas in probability and statistics makes studying these topics interesting and important. Many of the Internet sites in the book were used in assignments or brought to the author's attention by students in one semester (14 week) classes at Harvard University and one quarter (10 week) classes at Stanford University and The University of Chicago. High school algebra was sufficient background for all but a couple of topics in these courses.

The range of potential users is actually larger than instructors and students in high school and college introductory courses. Ideas of probability and statistics are being introduced to younger and younger audiences. Many middle school mathematics curricula contain units on describing data and understanding chance phenomena. Elementary, yet insightful, on-line demonstrations can be used in conjunction with traditional approaches.

In the other direction in terms of educational level, the importance of being able to communicate and relate theory to practice means that post-calculus statistics and probability classes can be enhanced by examining real-life applications of statistics. Often students in fields such as biology, economics, political science, psychology, and sociology enroll in classes focusing on statistical methods and the analysis of data. Links to examples and demonstrations can be used to supplement some post-calculus and applied methods courses. A general or continuing-education audience wanting a nontechnical, interesting introduction to probability and statistics might also find this book useful as a source of motivating examples.

How to use the book

The book's coverage of introductory statistics concepts is basically comprehensive. It is intended as a weekly or more frequent supplement to a longer textbook. This book provides current examples, on-line demonstrations, and questions that show the importance of topics in the course. The book can be used in a one quarter, a one semester, or a one year course and should be appropriate for a high school or college class. A textbook used in conjuction with this book will discuss practical implementation, such as details on making histograms. This book provides interesting examples and questions concerning interpretation to enhance understanding. The approach of this book is succinct. After a brief review and summary of a topic, the book presents Internet sites and study questions. Problems are of three types. The first helps teachers and students read and access material at a site. The second asks students to write about the site and the concepts. The third has the student perform computations. Communication is essential in applied statistics, and the practice of communication is emphasized throughout the book.

Instructors can be use this book for teaching and assignments. Many of the exercises can be translated into handouts or in-class activities. Others could be incorporated into homework assignments. Students can use the book to guide review and study. The material, because it is coordinated with an Internet site, can be used for instruction in a computer lab to show interesting on-line examples and simulations.

The book gives instructions for gathering data sets from on-line sources. The pedagogical value of interacting with data cannot be over emphasized. Students, on some sites, can enter their own data. Graphics are important in descriptive statistics, in describing simulation results, and for assessing the appropriateness of models. The book references several on-line illustrations and web sites that let students make graphic depictions of their own data. Several examples in the book reference articles available through Thomson Learning™'s INFOTRAC Online Library. INFOTRAC index numerous current major journals covering a variety of topics. The articles describe data, research designs, and statistical results. The book identifies articles appropriate for the intended audience and gives instructions for locating them, thereby helping the reader learn to use INFOTRAC effectively for research and study purposes.

Outline

The book is organized into ten chapters that correspond to topics found in most introductory statistics text-books and an eleventh chapter containing case studies. The first three chapters present Internet activities related to methods of describing data. Chapter one examples involve numerical and graphical summaries of univariate quantitative data through measures of center, spread, and shape. Chapter two focuses on descriptions of bivariate quantitative data: scatterplots, the correlation coefficient for measuring the degree of linear association, and the least-squares regression equation for predicting one variable in terms of another. These two chapters provide some examples of the use of transformations. Chapter three concentrates on summaries of categorical data and relationships between categorical variables. Chapter four concerns research designs and methods of data collection. Examples of experiments, surveys, including opinion polls, and observational studies with on-line documentation are discussed.

Chapters five through seven focus on probability, random variables and their probability models, and sampling distributions. Chapter five illustrates probability rules using Venn diagrams, tree diagrams, and tables. The chapter includes on-line demonstrations of popular, but challenging, probability questions. Chapter six presents problems on probability distributions that correspond to common measurement situations, including the binomial, Poisson, and normal models. Sampling distributions and two major results, the law of large numbers and the central limit theorem, are the topics of chapter seven. Chapters eight through ten address inferential statistics, the process of generalizing from a sample to a population. Chapter eight discusses confidence intervals, whereas chapter nine presents examples of hypothesis tests. The standard data structures and inferential procedures presented in introductory statistics are illustrated and contrasted. Chapter ten discusses inference for regression coefficients and some nonparametric statistical methods.

The book is concluded with a chapter of case studies. The reader uses ideas presented in other chapters to examine several dimensions of one research question or topic for each case study. A major challenge facing an applied statistician is deciding which descriptive or inferential procedure is appropriate. Research questions are studied from many perspectives, using multiple statistical techniques. Students and teachers should gain insight into the need for creative approaches to research questions.

The integration of the book with an Internet site means that links will be updated as necessary and new material will be added over time. These features increase the convenience and flexibility for integrating the Internet with teaching and learning statistics. The examples and links can be further explored for more in-depth and personalized study. Thus, the paper version of this book is just the beginning. Enjoy!

– Michael Larsen, April 14, 2003

P.S. I especially would like to thank Alice for her love and support and for providing links to several illustrative and stimulating web sites.

Chapter 1

Measurements on One Quantitative Variable

Introduction

The topic of descriptive statistics is concerned with methods of summarizing and displaying informa-tion. Effectively describing the data helps the reader understand what the data measure, look for patterns and anomalies, compare groups, and study relationships. It also is useful for demonstrating the validity or false-ness of a statement. Introductory statistics courses present numerical and graphical descriptive techniques for one or more quantitative and qualitative variables. Numerical summaries for a single quantitative variable include measures of the center and the spread of a distribution of values. The two most common measures of center are the mean ($\bar{x} = \frac{1}{n}\sum_{i=1}^{n} x_i$) and the median (the 50th percentile of the values $x_1, x_2, \ldots, x_n$). The spread of a set of values typically is described by the standard deviation ($s = \sqrt{\sum_{i=1}^{n}(x_i - \bar{x})^2/(n-1)}$) and the Interquartile Range (IQR = 75th percentile, or 3rd quartile, minus the 25th percentile, or 1st quartile).

There are several graphical methods for displaying the distribution of a single set of values. These include histograms, box plots, stem-and-leaf plots, and dot plots. The pictures are useful for visualizing the shape of the distribution of the values. Some distributions are symmetric, whereas others are skewed to the left or to the right. Some distributions have one primary peak, but others have multiple clusters. A few values might be outliers, that is, located somewhat far away from the vast majority of the data points; these might be apparent with some graphical methods, but not with others. One particular model of a distribution of values is given by the normal distribution. Both graphical and numerical summaries can be used to compare the data to the expected characteristics of a normal distribution. Normal probability plots, also called normal quantile or normal percentile plots, are scatterplots (see chapter two) used to assess the appropriateness of a normal model for a set of measurements.

Remarks

Descriptive statistics can be very useful and interesting when used to compare groups. Do two or more groups have similar distributions? Does the same relationship between variables apply to observations from two groups? Are the proportion of respondents in certain categories the same across various groups? In writing about data, it is critical to comment on the context. What do the variables measure? What are the units of measurement? What is the purpose of collecting the data? It is important to read the question and understand what it is asking before producing numerical and graphical summaries. Answers should not be automatic and based solely on first impressions, because several questions can be asked about each dataset.

It is important to understand the relationship between different summaries. Several examples demonstrate the meaning of this statement. For strongly right-skewed distributions (the right tail of the distribution is longer than the left), the mean will be larger than the median. A few extreme outliers can increase the standard deviation a lot. Pairs of values represented in a scatterplot can be used to construct histograms of both variables. Transformations of the original measurements, especially in the case of quantitative variables, often are useful and affect numerical and graphical summaries. The effect of linear transformations, such as Fahrenheit to Celsius or Inches to Centimeters, are the easiest to describe. The impact of a nonlinear transformation, such as square root (often useful for measurements of areas), cube root (often useful for measurements of volumes), or log (useful in some cases where ranges are extreme), is harder to predict.

Finally, descriptive statistics, and applied statistics in general, is a creative activity. For every research question, there are numerous variables that might be of interest. For every variable, there are a variety of ways to record information. For each set of measurements, there are a number of useful descriptive summaries. New, innovative displays appear in newspapers and research journals all the time.

Outline

Section 1.1 presents examples that compare and contrast measures of the center of a set of values. Section 1.2 focuses on measurements of spread. Section 1.3 concerns the shape of distributions of data. Examples of pie charts are given in chapter 5 on probability. Questions concerning general percentiles are located in chapters 5 and 6 on probability and distributions of random variables. The Companion Website can be accessed at http://larsen.duxbury.com. Please enter the Serial Number from the back cover when prompted.

1.1   Measurements of the Center

The following table outlines the problems contained in section 1.1.

Problem	Description
1.1 (A)	Mean and median career records of athletes in various sports.
1.1 (B)	Mean and median heights and weights of rowers and coxswains.
1.1 (C)	On-line demonstration of the impact of unusual observations on means and medians.
1.1 (D)	Mean and median wage data by occupation from the Bureau of Labor Statistics.
1.1 (E)	The definition of time, errors in measurements of time, and average errors of various clocks.
1.1 (F)	The mean and median size of Cuckoo eggs laid in the nests of other birds.

Problem 1.1 (A)

Sports statistics: the mean and the median as descriptions of the center

Measurements of the performance of sports players are regularly collected and used to describe their performance. In this activity you will record career information on a few players from sports web sites, compute means and medians, and examine whether or not these summaries accurately describe the center of the data. Below are listed a few sports web sites with suggestions for measurements to record. Record the career records in at least two categories for at least two players, then answer the questions below.

- Baseball: http://www.mlb.com. Select "History" from index across the top of the page, then select "Historical Stats" from the index on the left of the page, and then enter a player's name. Historical players that were famous for their ability to hit the baseball include Babe Ruth, Hank Aaron, Roger Maris, Ted Williams, and Joe DiMaggio. Recent and active players compared to this group include Mark McGuire, Sammy Sosa, Ken Griffey, Jr., and Barry Bonds, Jr. For these players season statistics of interest include number of home runs, number of batting attempts ("at bats"), and batting average.

- Women's Golf: http://sportsillustrated.cnn.com/golfonline/golfstats/lpgaSearch.html. Enter a player's name. See http://sportsillustrated.cnn.com/golfonline/tours/whoshot for names of some players. An alternative to comparing players is to compare tournaments. From http://sportsillustrated.cnn.com/golfonline/women, click on Results, then a tournament, then a year. "Score" gives the overall score as the sum of three rounds. "To Par" gives the number of strokes below par. "Dollars" gives the amount the player won.

- Men's Soccer: http://www.mlsnet.com/archive or http://www.mlsnet.com/archive/register_players.html. Select a few players that have played at least 4 seasons. Season statistics of interest include games played (GP), games started (GS), minutes played (Min), goals scored (G), goals attempted (A), and shots on goal (SOG). For players with at least one shot on goal per season, you can divide goals scored by shots on goal to get a rate (goals per attempt) for each season.

- Women's Basketball: From http://www.wnba.com, select "statistics." The names of league leaders appear in tables on this page. Record the names of several players. Go to http://www.wnba.com/players. Click on the link to your chosen players and record her information for categories such as number of games, field goals, free throws, points, and points per game. If the player has not played at least three seasons, then select a different player.

1. Calculate the mean and median of measurements recorded for a few players. Describe what these summaries represent. Do the statistics indicate that one player is better than another in terms of this measure of performance?

2. Are any data points far from the mean and median?

3. Delete one or more unusual points for a player and recalculate the mean and median. Unusual points might be very high or very low in comparison to the other points. They might appear at the beginning or at the end of a player's career or pertain to a season in which the player was injured and missed several games or tournaments. How are the means and medians affected by the deletion of these points? Be sure to explain why you choose the points you decided to delete.

4. See http://www.shodor.org/interactivate/activities/stemleaf on stem-and-leaf plots. Click on the "what?" button and read the summary. Enter the data into the data window for a single player.

Problem 1.1 (B)

Weights of Rowers versus Coxswains: the impact of extreme observations on means and medians

The Data and Story Library (DASL) illustrates the impact of an outlying observation on the mean and median measures of center by considering weights of Rowers and Coxswains: http://lib.stat.cmu.edu/DASL/Stories/Rowersvs.Coxswain.html.

1. Compute the mean and median for the Cambridge and Oxford teams separately. The data are located at http://lib.stat.cmu.edu/DASL/Datafiles/Crews.html. What impact does including or excluding the coxswain have on the measures of the center? *Comment*: Often in statistical analyses it is useful to separate subjects according to a variable that divides them into distinct groups, such as female/male or coxswain/rower, and analyze them separately. In this application, however, the total weight of the boat is relevant to speed.

2. Look at the team roster for a heavy weight college crew team. For example, the athletics department at Harvard College is located at http://www.athletics.harvard.edu (or http://gocrimson.ocsn.com). Select the sport "M - Crew HWT" for Men's Heavy Weight Crew and click on the Roster to the right. The direct links is http://gocrimson.ocsn.com/sports/m-crewhvy/mtt/harv-m-crewhvy-mtt.html.

   (a) What is the average weight of the members of the team? What is the median?

   (b) Can you identify the coxswains? Calculate the average weight of the coxswains and rowers, separately. Calculate the medians. How do they compare?

   (c) How do the groups and summaries of the center compare in terms of height?

   (d) Repeat the exercise above for groups and schools that interest you. Here are some examples with data on the Internet.

      • Weights of Harvard's Men's Lightweight Crew:
        http://gocrimson.ocsn.com/sports/m-crewlt/mtt/harv-m-crewlt-mtt.html.
      • Heights of Yale's Women's Crew: http://www.yale.edu/athletics, http://www.yale.edu/rowing, http://www.yale.edu/rowing/wom_roster.html. You may assume for the purpose of this analysis that the shortest women are coxswains. This might not, however, be 100% correct.
      • Height and Weight of University of California, San Diego's men's crew team: http://crew.ucsd.edu/vroster.html.
      • Height of Johns Hopkins University crew teams. The web page separates coxswains from rowers. In order to analyze a full boat and see the impact of the lighter, smaller coxswains, it is necessary to combine the data sources. http://www.jhu.edu/~jhucrew

Problem 1.1 (C)

Demonstrations of the outlier impact on means and medians.
Two applets below illustrate the impact of single points on means and medians.

1. See http://statman.stat.sc.edu/~west/applets/box.html. Move the point around. Describe in your own words what impact the movable point has on the mean versus median calculation.

2. See http://www.ruf.rice.edu/~lane/stat_sim/descriptive/index.html. Note the values of the mean and median. In this exercise, you can also note the standard deviation (SD) of the points. In the next section, you will be asked how the SD changes.

   (a) Add a point or two to the right side of the histogram. How does this affect the mean and median?

   (b) Remove these points (click on the squares you added) and add new points to the right side of the histogram. How does this affect the mean and median?

   (c) Make a uniform distribution. A uniform distribution has an equal number of observations at each observed value. What are the mean and median?

(d) Make a distribution that takes values only at 3 and 7. Where is the center? Make a distribution that takes values only at 2 and 8. Where is the center?

Alternative applet: http://www.shodor.org/interactivate/activities/plop.

Problem 1.1 (D)

Bureau of Labor Statistics Wage and Salary data by state and occupation

The Bureau of Labor Statistics (http://www.bls.gov) is the official source of wage data for the U.S.: http://www.bls.gov/bls/blswage.htm. Select your state: http://www.bls.gov/oes/2001/oessrcst.htm.

1. For all occupations, which is higher, the mean or median hourly wage? What does this imply about the distribution of wages, is it symmetric, skew left (long left tail), or skew right (long right tail)?

2. What is the relationship between mean hourly wage and mean annual earnings? Divide the mean annual amount by the mean hourly amount. Convert this number to weeks (divide by 40 hours for a standard week). Does this make sense?

3. Find an occupation for which the median hourly wage is less than the mean hourly wage. Is this an occupation classification for a lot of people in your state? Is the center of the wage distribution lower or higher than the overall center?

Problem 1.1 (E)

Errors in measuring time: Huygens clocks and a small experiment

The National Institute of Standards and Technology (NIST; http://www.nist.gov) is responsible for measurement standards in the U.S. government. One task of NIST is to keep track of time (http://nist.time.gov, http://www.boulder.nist.gov/timefreq/index.html). For a discussion of accuracy of time keeping, see the article "Accurate Measurement of Time" by Wayne M. Itano and Norman F. Ramsey in the July, 1993, issue of *Scientific American*: http://www.boulder.nist.gov/timefreq/general/generalpubs.htm.

1. Define three quality standards for time measuring instruments or methods: stability, reproducibility, and accuracy. See page 56 of the Itano and Ramsey article. Comment on the stability, reproducibility, and accuracy of the quartz-crystal oscillator clock of Horton and Marrison (see page 57).

2. Christian Huygens produced the first pendulum clock December, 1656. See http://www.sciencemuseum.org.uk/on-line/huygens.

    (a) Huygens' first clock was accurate to within 10 minutes per day. Report ten imaginary measurements of the length of the day that average 10 minutes too long. Remember, a day is 1440 minutes long. What is the mean and median of your measurements? Although its best to use real data when possible, it also is important to be able to be creative and imagine what values could be recorded.

    (b) Huygens' second clock was accurate to within 1 minute per day. Report nine imaginary measurements of the length of the day that average 1 minute too long. What is the mean and median of your measurements?

    (c) How do stem-and-leaf plots of your two data sets compare? See problem 1.1 (A) for the stem-and-leaf plotter.

3. How good is your sense of time? Try to guess how long 10, 20, and 40 second durations of time are. Try to guess each duration a few times. This can be done with a partner or in a group. See http://www.shodor.org/interactivate/activities/stopwatch for a stopwatch.

   (a) Calculate averages and medians of your data and data from others for the three lengths of time. Which length of time is the hardest to guess? That is, which had the largest errors? Are there any very unusual points?

   (b) If you have several observations at each time period, describe the shape of the distribution.

   (c) Subtract the actual time (10, 20, or 40 seconds) from the estimated time. What are the averages and medians of these data? How would stem-and-leaf plots of these data appear?

Problem 1.1 (F)

Those tricky cuckoos: mean and median egg size by host species

Cuckoo birds (see http://lib.stat.cmu.edu/DASL/Stories/cuckoo.html) have a habit of laying their eggs in nests of other birds. Here are some egg measurements: http://lib.stat.cmu.edu/DASL/Datafiles/cuckoodat.html.

1. Based on the pictures, rank the host birds in terms of median cuckoo egg size. You may report ties if a pair is too close to call.

2. Based on the pictures, rank the host birds in terms of mean cuckoo egg size.

3. Compute the summaries for the different groups to check your answers.

1.2   Measurements of Spread

The table below outlines the problems of section 1.2.

Problem	Description
1.2 (A)	The interquartile range (IQR) and standard deviation (SD) of career sports records.
1.2 (B)	IQRs and SDs of heights and weights of college rowing team members.
1.2 (C)	An on-line demonstrations of the impact of unusual observations on IQRs and SDs.
1.2 (D)	Data on dogs, zoo animals, and fish are used to illustrate summaries of center and spread.
1.2 (E)	The medians and quartiles of salaries for statisticians at various academic levels.
1.2 (F)	The spread of time measurements is related to measurement error and bias.
1.2 (G)	The IQR and SD of Cuckoo egg sizes are compared by host species.

Problem 1.2 (A)

The standard deviation and interquartile range as descriptions of spread of career sports records

In addition to the center, measurements of the performance of sports players can summarized by their spread. Standard deviations can be computed for two or more measurements. Interquartile ranges (IQR) are meaningful for eight or more observations. Refer to the data collected in problem 1.1 (A).

1. Calculate the standard deviation and IQR for a few players. Describe what these summaries represent.

2. How do the players compare in terms of spread? Is one player more consistent than another?

3. Delete one or more unusual points for a player and recalculate the standard deviations and IQRs. What impact do the extreme points have on these measures.

Problem 1.2 (B)

Weights of rowers versus coxswains and the sensitivity of measures of spread

See the DASL site concerning Rowers and Coxswains:
http://lib.stat.cmu.edu/DASL/Stories/Rowersvs.Coxswain.html.

1. Compute the standard deviations of the weights for the for the Cambridge and Oxford teams with and without the coxswains' data. What impact does the outlier make on the standard deviation?

2. Return to the team roster data you collected in problem 1.1 (B).

    (a) Compute the standard deviation and IQR for heights and/or weights using all the observations.

    (b) Compute the standard deviation and IQR for heights and/or weights using only the rowers.

    (c) Comment on the impact outlying observations have on SDs and IQRs.

Problem 1.2 (C)

On-line demonstration of the outlier impact on summaries of spread

The applet below illustrates the impact of single points on measures of center and spread. See
http://www.ruf.rice.edu/~lane/stat_sim/descriptive/index.html.

1. What are the standard deviation and interquartile range for the 9 observations given? Add a point at 5. What are the SD and IQR for these 10 observations?

2. Remove the added point at 5 (click at the base of the top block above 5) and add three points at 9. Now there are 12 points. How do the SD and IQR change?

3. Add three more points at 1. How do they affect the SD and IQR?

Problem 1.2 (D)

Measurements of center and spread for some animals and fish

Measurements on several types of animals are available on-line. A few sites are given below. For one or more of the sets of measurements, answer the questions below.

- Dogs: see http://sln.fi.edu/school/math3/dogs.html. Imagine that you have one dog of each species and that the dog of each species has height, weight, and a life span equal to the midpoint of the given interval. If you have a dog, you can include its height and, if possible, weight measurements.

- A cross-species comparison: see http://www.zooregon.org/Cards/cardindex.htm. Select 12 to 20 cards and record information from them. Most measurements are given in ranges, so you'll have to select a representative value. Be careful with units of measurements! Grizzly bears are measured in feet (') and pounds (lbs.), whereas Egyptian fruit bats are recorded in inches (", 1 foot = 12 inches) and ounces (oz., 1 pound = 16 ounces).

- Fish: The Department of Wildlife and Parks, State of Kansas (http://www.kdwp.state.ks.us), records the size of the largest fish of various species caught in Kansas (http://www.kdwp.state.ks.us/fishing/Fish_Records.html). Pick at least 8 of the record species. Convert length to inches and weight to ounces. You can also use date of catch: record the year of the catch or the number of years since 1900, e.g., 1977 equals 77 and 2003 equals 103. If you interested in your state, searching using Google (http://google.com) on the name of your state and the "fishing records."

1. Calculate means and medians of your measurements. Which is more representative of the center?

2. Calculate SDs and IQRs for your data.

3. Are there any extreme observations? Which of your calculations are most affected?

4. Write a paragraph expressing what these numerical summaries tell you about the animals.

Problem 1.2 (E)

Medians and quartiles of academic salaries of statisticians by rank and experience

The American Statistical Association (http://www.amstat.org) collects information on the salaries of academic statisticians and biostatisticians (statisticians working in biology, medicine, and public health). See http://www.amstat.org/profession/index.html#salary.

1. The link "2002-2003 Academic Statisticians" and "2001-2002 Academic Statisticians" give summary salary information for academic statisticians for whom responses were received. There likely were some reporting biases, but that is another issue. Contrast the assistant, associate, and full professors at research universities in terms of center and spread of salary distributions. Make this comparison separately for the liberal arts colleges and the research institutions.

2. Contrast the assistant professors at the liberal arts colleges and at the research institutions.

3. What impact does the years of experience have on the center and the spread of salary distributions?

4. The link "2002 Biostatisticians and Biomedical Statisticians" provides salary information on these groups. How has the salary distribution changed over time for the assistant and full professors?

Problem 1.2 (F)

Errors in measuring time: measurement error and spread

The role of the National Institute of Standards and Technology (NIST; http://www.nist.gov) for setting time standards was discussed in problem 1.1 (E). See again the article "Accurate Measurement of Time" by Wayne M. Itano and Norman F. Ramsey in the July, 1993, issue of *Scientific American*: http://www.boulder.nist.gov/timefreq/general/generalpubs.htm.

1. Two of the three quality standards for time determination, stability, reproducibility, and accuracy, are related to the spread of time measurements. The other is related to measurements of the center. Which are related to spread? Briefly explain how they are related.

2. Consider the artificial values you made up to represent hypothetical measurements on Huygens' clocks.

   (a) Calculate SDs and IQRs for the two data sets.

(b) Why does one have larger measurements of spread?

3. How good is your sense of time? Which set of measurements you generated in problem 1.1 (E) has the greatest spread? the least?

4. Subtract the actual time (10, 20, or 40 seconds) from the estimated time in your experiments. How do the measurements of spread on the original data correspond to measurements of spread for these data?

Problem 1.2 (G)

Those tricky cuckoos: comparison of spread of egg size by host species

See http://lib.stat.cmu.edu/DASL/Stories/cuckoo.html and problem 1.1 (F) on egg-laying by Cuckoo birds.

1. Based on the pictures, rank the host birds in terms of interquartile range.

2. Based on the pictures, rank the host birds in terms of standard deviation.

3. Compute the summaries for the different groups and check your answers.

1.3   Graphical Depictions of Shape

Two graphical methods for describing the shape of a distribution of data appeared in the previous sections: stem-and-leaf plots and histograms. This section presents bar graphs, dot plots, and box plots in addition to a more thorough study of histograms.

Problem	Description
1.3 (A)	Bar graphs about the Smithsonian Institution budget and women in the movie industry.
1.3 (B)	Side-by-side box plots for comparing various distributions and sports careers.
1.3 (C)	Side-by-side box plots and classification of 3 species of flea beetle.
1.3 (D)	A histogram applet with several datasets and animal size data: the impact of bin size.
1.3 (E)	Salary data for faculty at top universities in the 1990s: relationships among summaries.
1.3 (F)	Comparison of results of an on-line response time experiment involving two conditions.
1.3 (G)	Contrasts of graphical summaries of data from a reading activities experiment.

Problem 1.3 (A)

Bar Graphs: the Smithsonian Institution and Women Movie Directors

Bar graphs are useful for displaying the frequency of responses or amounts in two or more categories. Go to http://www.shodor.org/interactivate/activities/bargraph.

1. Look at the three data sets and their bar graphs available on this page. Describe briefly what is depicted by the three graphs.

2. The Smithsonian Institution (http://www.si.edu) preserves and presents the cultural, economic, political, and scientific history of the U.S. Information on its federal budget can be found at http://www.si.edu/about/budget. Click on "07 - Bar Chart- Salaries & Expenses Summary by Function" and "08 - Bar Chart- Summary by Function." The charts for 2003 and 2004 represent the Smithsonian's requests to the U.S. Congress, whereas the 2002 chart is the actual budget.

(a) What is the total budget in 2002? An approximate answer is acceptable.

(b) How is the budget changing over time in the area of research, facilities, and IT (information technology)?

(c) How is the allocation of the budget changing over time?

3. Women have made great contributions to film and television. The web site "Movies Directed by Women" (http://www.moviesbywomen.com) is dedicated to promoting women's participation at all levels of movie making. Examine the graphs at the bottom of the page: http://www.moviesbywomen.com/stats2001.html. Note that the figures present percentages.

(a) According to figure 1, in 2001, was the percent of women employed in these important roles higher in the top 100 films or in the films ranked 101-250?

(b) In figure 1, which bar, the one for the top 100 films or the one for the top 250 films, for producers is based on more people? Why?

(c) According to figure 2, how has the percent of women employed in these important roles in the top 250 films changed in the past few years?

(d) According to figure 3, how has the percent of women employed in these important roles in the top 100 films changed in the past few years?

4. Return to the interactive bar graph on question 1. Click on the "What?" button. Look at the two examples on this page. The top one is a bar graph, whereas the lower one is a histogram. Describe how these are different. After you write your explanation, read the discussion at the "Histograms vs. Bar Graphs" link located at the bottom of the page. Revise your answer as you wish.

Problem 1.3 (B)

Side-by-side box plots: six examples and sports statistics

Box plots (see http://www.shodor.org/interactivate/activities/boxplot/index.html) present a five number summary (minimum, first quartile, median, third quartile, and maximum) graphically in a way that is effective for comparing groups.

1. Look at the "NBA Team Payrolls" example. The scale is presented in scientific notation. That is, $3.599e4 equals $ 3.599 \times 10^4 = $35,990,000. What are the median and interquartile range? What is the range? Is this distribution skewed or symmetric?

2. Are there any outliers according to the 1.5 IQR rule? The 1.5 IQR rule is a hueristic, or guideline, that identifies points as extreme if they are more than 1.5 IQRs above the third quartile or below the first quartile. Calculate the cutoff points and then look at the data in the data window. The web page will not show you this cutoff, but will show you individual points beyond the first and third quartiles: check the "Exclude extreme outlier in graph" box. Points near the quartiles are not really extreme, but individual points near the maximum and minimum sometimes are.

3. Look at another example. Describe in your own words what the box plot shows.

4. Click the cicle labelled "Graph by category," then look at "Amount Spent Per Student" and "Gas Mileage for 2000 cars by size."

(a) How are the categories defined in these two examples?

(b) How do the distributions compare in terms of center and spread?

5. Enter your data from problem 1.1 (A) and 1.2 (A) for two atheltes with at least 8 data points each. Make up unique codes for the names of your two atheletes, such as "b" for Babe Ruth and "h" for Hank Aaron. Follow each value by a comma and the code for the athlete. See the data window of the "Amount Spent Per Student" and "Gas Mileage for 2000 cars by size" for examples of the data format. Change the label for the data set.

   (a) How do the distributions compare in terms of center and spread?

   (b) Are the outliers according to the 1.5 IQR rule for either player?

   (c) How do the shapes of the distributions compare?

Problem 1.3 (C)

Side-by-side boxplots and classification of 3 species of flea beetle

Can you distinguish flee beetle species based on measurements of their size? The DASL site http://lib.stat.cmu.edu/DASL/Stories/FleaBeetles.html compares side-by-side boxplots of two measurements on three species. See also http://www.ipm.uiuc.edu/vegetables/insects/corn_flea_beetle and http://www.ento.psu.edu/extension/factsheets/flea_beetle.htm.

1. Describe differences among the three groups in averages and standard deviations of the two variables.

2. Which distributions are skew and which are symmetric?

3. Which have outliers?

4. Confirm the calculations represented by the pictures by finding a group's 1st, 2nd, and 3rd quartiles.

5. Using the 1.5 IQR rule, how extreme do beetles in these groups have to be to be considered outliers?

   What else is here?   Analysis of variance (ANOVA) is used to test whether or not there are differences between several groups. Multivariate ANOVA (MANOVA) is used to examine two or more variables simultaneously on several groups. The Scheffe' procedure is one of a number of procedures that are used in ANOVA to control the chance of errors overall when several comparisons are being made.

Problem 1.3 (D)

A histogram applet with several datasets and animal size data: the impact of bin size

Histograms summarize the data by reporting the number or percent of observations that appear in an interval. Stem-and-leaf plots drawn on their side (horizontally) resemble histograms.

1. See http://www.stat.sc.edu/~west/javahtml/Histogram.html. The data represented in this histogram are durations of eruptions of the Old Faithful geyser at Yellowstone National Park.

   (a) Move the binwidth indicator to 1.0. The fraction of observations in each interval along the horizontal axis is represented by the area of the rectangle, which in this case is equal to the height (area = base times height). Write down the four intervals and the fraction of observations in each interval.

   (b) Set the binwidth to 0.5 or as close as you can. The area still represents the proportion of observations in each interval. Write down the seven (or eight) intervals and the proportions.

2. Go to http://www.shodor.org/interactivate/activities/histogram. This site presents histograms of several data sets and allow you to manipulate the interval size, or bin width. Look at the six datasets.

   (a) For two data sets, what are the mean and standard deviation? Include units of measurement in your answer.

   (b) For two data sets, describe the shape of the distribution. Are they skew or symmetric? Are there outliers?

   (c) For one data set, make the bin width large, then medium, then small. Record the name of the data set, the bin widths, and how many intervals are used. How does the appearance of the distribution change as you change the bin widths? Do outliers appear in one picture but not in others?

3. Press the "Clear" button. Enter your data from problem 1.2 (D) into the data window, change the name of the data set, and press "Update Histogram." Adjust the bin width for your data.

   (a) Describe the shape of the distibution of your data.

   (b) How do the summary statistics of problem 1.2 (D) relate to the histogram?

4. Press the "Clear" button. Click on "More Data Sets" at the upper left of the page, then the link to the "Old Faithful" dataset. Hightligth the data and copy them into the data window on the applet page. Change the name in the "Describe your data" window and press the "Update Histogram" box.

   (a) Set the interval size to as close to 1 as you can. What are the endpoints of and frequencies in the intervals? Press "Show Frequency Table" to open a window. What fraction of the eruptions are in each interval?

   (b) Set the interval size to as close to 0.5 as you can. What are the endpoints of and frequencies in the intervals? Press "Show Frequency Table" to open a window. What proportion of the eruptions are in each interval?

Problem 1.3 (E)

Summaries of Faculty Salaries at Top Universities in the 1990s

The average salaries for assistant, associate, and full professors at 50 universities are displayed at http://lib.stat.cmu.edu/DASL/Stories/Professors'Pay.html and the data are reported at http://lib.stat.cmu.edu/DASL/Datafiles/FacultySalaries.html.

1. Based on looking at the side-by-side dot plots, how do the means of the average salaries compare across the levels of professors?

2. How do the standard deviations of the average salaries compare?

3. If you made histograms out of the data, which group would have the most skew histogram? You can make the histograms if you want to check your answer.

4. Compare salaries by professorial level at the CIC institutions to those at other institutions in the data base. Make appropriate numerical and graphical comparisons.

Problem 1.3 (F)

A graphical and numerical comparison of on-line response time experiment results

The web site http://www.ruf.rice.edu/~lane/stat_sim/compare_dist/index.html has you perform a timed dexterity test. The times to click the mouse pointer on a square is recorded. The squares are either large or small. Perform this activity. Record results and draw pictures of the graphs.

1. Compare the centers of the distributions. Write full sentences for your answer.

2. Compare the spreads of the distributions. Write full sentences for your answer.

3. Do the distributions have the same shapes? Does either distribution of values have unusal points?

   For your information: The trimmed mean is the mean of all but the most extreme observations. The sem is the standard error of the mean, $s/\sqrt{n}$. The coefficient of skewness is a numerical summary of the degree of skewness. Its positive or negative for distributions that are skewed to the right or left, respectively. The coefficient of kurtosis is a numerical summary of the degree to which the distribution has both extremely large and small values.

Problem 1.3 (G)

Graphical comparisons of the impact on reading ability of directed reading activities

DASL reports on an experiment concerning the impact of directed reading activities on reading ability of third grade students. See http://lib.stat.cmu.edu/DASL/Stories/ImprovingReadingAbility.html and http://lib.stat.cmu.edu/DASL/Datafiles/DRPScores.html.

1. The graph shows side-by-side dot plots of the data for the treatment group, which participated in the directed reading activities, and the control group, which did not. Compare and contrast using the pictures the distributions of scores for the two groups.

2. Make side-by-side box plots of the data for these two groups using the data. What do the dot plots show you that the box plots do not?

3. Will histograms for these two groups of scores be skew to the left, symmetric, or skew to the right? Will they be similar or different in shape? State an answer and draw a rough sketch before you make histograms.

   What else is here?   The site reports the summaries of a two-sample t-test for assessing whether the difference in means is due to chance variability or to an actual difference in average reading scores. Methods of statistical inference in chapter 9 can be used to reproduce these numbers.

1.4   Transformation of a Single Quantitative Variable

Problem	Description
1.4 (A)	Linear transformations of temperatures and windspeeds.
1.4 (B)	Annual revenue of Fortune 500 companies and the logarithm transformation.
1.4 (C)	Measurements of the speed of light and the acceleration due to gravity.

Problem 1.4 (A)

The effect of a linear transformation on summaries of temperature and windspeed

The current weather in several Illinois cities is available on-line from the Chicago Tribune: http://weather.chicagotribune.com/US/IL. Pick at least 12 cities and record the weather (in degrees Fahrenheit) and wind speed (in miles per hour). You can change the state to your state by replacing "IL" with your state's abbreviation.

1. Summarize the temperature data. Make a graphical display, including labels of axes.

2. Convert the temperature measurements to the Celcius scale: Celcius = (Fahrenheit - 32) (5/9). Summarize the temperature data and make a graphical display. How do results compare to those on the Fahrenheit scale?

3. Go to http://www.cnn.com/WEATHER. Click on the temperature link beside a map of a region of the world other than North America. Temperatures on the map are in Celcius. Summarize the temperature data. Make a graphical display, including labels of axes.

4. Convert the temperature measurements to the Fahrenheit scale, summarize the values, and compare to the previous part: Fahrenheit = 32 + Celcius (9/5).

5. Can you convert the miles per hour wind data to kilometers per hour? How does this change affect measures of center and spread and graphical displays? A search on "metric conversion tools" yields numerous sites, such as http://www.convert-me.com/en.

   More Information: See http://www.temperatures.com/scales.html on temperature scales.

Problem 1.4 (B)

The effect of a logarithmic transformations on sales of Fortune 500 companies

Go to http://www.fortune.com/fortune/alllists to see the lists of lists provided by *Fortune* magazine. Data that have great ranges sometimes are usefully transformed to a logarithmic scale. Click on the link to the "Fortune 500" companies. Pick a number between 1 and 10 inclusive. Record the sales of every thirtieth company beginning with the number of the company you selected. For example, if you choose 2, record sales for companies 2, 32, 62, 92, etcetera. Record sales nine or more companies.

1. Compute the mean and standard deviation. You can compute these on the millions of dollars scale or in the dollars scale, and these summaries will be on the same scale. As an example of the conversion, $20,000 in this table means $20,000,000,000.

2. Take the logarithms to the base 10 (log or log10, depending on the calculator) of these numbers. Be careful of the scale here: $\log_{10} 10,000 = 4$ but $\log_{10} 10,000,000,000 = 10$. Compute the mean and standard deviation on the log scale. Is the mean equal to the log of the mean on the original scale? Is the SD equal to the log of the SD on the original scale?

3. Make a graphical summary of your data on both scales (log and non log). How do the shapes of the distributions compare?

4. Find the first quartile, median, and third quartile on the original and on the log scale. How are they related?

14

Problem 1.4 (C)

Speed of Light: a nonlinear transformation and summaries of one variable

DASL reports a story (http://lib.stat.cmu.edu/DASL/Stories/EstimatingtheSpeedofLight.html) about measuring the speed of light. The measurements (http://lib.stat.cmu.edu/DASL/Datafiles/SpeedofLight.html) are reported as deviations from 24,800 nanoseconds; the speed of light is, of course, positive.

1. Which is higher, the mean or the median? What are the standard deviation and the interquartile range? Calculate these numerical summaries on the reported scale.

2. What are the mean and median on the original (nanosecond) scale? What about the standard deviation and interquartile range?

3. What happens to the measures of center and spread when you drop the lowest observation? the lowest two observations? Describe your results in a sentence or two.

4. Imagine trying to repeat Simon Newcomb's experiment. Light travels so fast that you could not realistically do it without very sophisticated equipment. You can, however, experience some of the challenges of making precise measurements by designing and implementing a study to measure gravity. Hold a ball at a height of one meter. Drop the ball and record the time it falls until it hits the ground. A stop watch with hundredths or thousandths of a second would be useful
(see http://www.peoplessavingsonline.com/justforkids).

   (a) How accurate is your measurement of the starting height of the ball? How accurate is your measurement of the time from when it is released to the time it hits the ground? Write a few sentences describing your experiences.

   (b) Take several measurements. Report the average and median time. Report the SD and IQR. Are there any unusual data points? What accounts for their difference from the others?

   (c) The relationship of time to gravity is given by *distance* = (*acceleration due to gravity*)(*time*)^2/2, so *time* = $\sqrt{2/(acceleration\ due\ to\ gravity)}$. Look up the value of gravity on the National Institute of Standards and Technology (NIST): http://physics.nist.gov/cuu/Constants/index.html, search for "gravity." Convert this number to time in seconds. How close are your measurements?

   (d) Convert your measurements into estimates of acceleration due to gravity:
   (*acceleration due to gravity*) = 2*distance*/(*time*)^2. Calculate the measures of center and spread. How close is the center to the official value? The units for these values of meter per sec^2.

More information: Read more about measuring the speed of light at
http://www.what-is-the-speed-of-light.com and Simon Newcomb at
http://www-groups.dcs.st-and.ac.uk/~history/Mathematicians/Newcomb.html.

Chapter 2

Measurements on Two Quantitative Variables

Introduction

Pairs of values measured on two quantitative variables (both variables recorded for each subject) can be displayed in a scatterplot. Scatterplots show direction (positive, negative, or nonexistent) and strength (weak, moderate, or strong) of associations and other patterns. They can be used to identify observations with unusual values on one or both variables and clusters of observations. If one of the variables records time, a graph with time on the horizontal axis is a time series plot. Time series can display serial correlation and cyclical trends in addition to the other features of scatterplots.

The strength of the linear relationship between two quantitative variables can be summarized by the correlation coefficient $r = \frac{1}{n-1} \sum_{i=1}^{n} \frac{(x_i - \bar{x})}{s_x} \frac{(y_i - \bar{y})}{s_y}$, where s_x and s_y are the standard deviations of the $x-$ and $y-$ values, respectively. If one wants to think of predicting one set of values $(y_1, y_2, \ldots, y_n)$ based on the other set of values (the x's), then one option is the least-squares linear regression equation. The best equation, according to the least-squares criterion, is given by $\hat{y} = a + bx$, where $b = \dfrac{\sum_{i=1}^{n}(x_i - \bar{x})(y_i - \bar{y})}{\sqrt{\sum_{i=1}^{n}(x_i - \bar{x})^2}\sqrt{\sum_{i=1}^{n}(y_i - \bar{y})^2}} = r\frac{s_y}{s_x}$ and $a = \bar{y} - b\bar{x}$. The prediction error, or residual, for case i is $e_i = y_i - \hat{y}$.

Remarks

Graphical methods in general reflect the instrument used to record measurements through rounding and scale. Thus, it is important to pay attention to the units of measurement for both variables. Graphs should have labels and clear, accurate axes. The appropriateness of the numerical summaries of correlation and the linear prediction equation can be evaluated by examining the scatterplot of y versus x. Another useful plot is a scatterplot of the residuals (e's) versus the x's. In situations where there is slight curvature of the relationship, the curvature and large outliers often are more apparent in the residual plot. If data have been collected on different groups of subjects, consider scatterplots and summaries of the groups separately as well as combined together. Results will not necessarily be the same.

Methods for summarizing two quantitative variables should not be used to the exclusion of methods presented in chapter 1. A scatterplot depicts the individual distributions of values of both variables, but not as clearly as univariate methods. Cases that are outliers in one dimension are likely to be important in two dimensions as well. Finally, strong association does not mean causation. Size, for example, often explains strong associations between several variables pairs. Increasing the magnitude of one variable for a small unit will not necessary cause an increase in the other variables.

Outline

Section 2.1 presents examples of scatterplots and correlation. Section 2.2 contains problems on linear regression. Section 2.3 is concerned with transformations of two quantitative variables and the impact on bivariate relationships. Section 2.4 illustrates issues involved with time series measurements. Considerations important in making statements of causation are discussed in chapter 4. The bivariate normal distribution appears in section 6.6.

2.1 Scatterplots and Correlation

The following table outlines the problems contained in section 2.1. Directions for gathering other data sets can be found in problems 10.2 (A), 10.2 (B), and 11.2.

Problem	Description
2.1 (A)	Scatterplots of the percent vote by state in several years for president of the United States.
2.1 (B)	Correlations between measurements on animals.
2.1 (C)	An on-line applet for making scatterplots with specified correlations.
2.1 (D)	Correlations between sets of stock values at two time points.
2.1 (E)	Education spending by state and correlation, scatterplots, and causation.
2.1 (F)	Use two sets of measurements on Flea beetles to distinguish between species.
2.1 (G)	Readings on the ecological fallacy.

Problem 2.1 (A)

Voting for the president of the United States: scatterplots and clusters

The scatterplot on http://lib.stat.cmu.edu/DASL/Stories/VotingforthePresident.html displays the percent voting by state for Carter (1980, horizontal axis) and Mondale (1984, vertical axis) for U.S. president.

1. In your own words, explain what is depicted in the figure. What do the highlighted marks represent? Identify the states that have these values. The data: http://lib.stat.cmu.edu/DASL/Datafiles/Votes.html.

2. Describe the association between the percent state vote for Carter and Mondale. Explain why the association is so strongly positive. For U.S. maps, see http://nationalatlas.gov/printable.html, select "printable maps" next to Presidential Elections 1789-2000 (http://nationalatlas.gov/electionsprint.html), then the "Preview Map" for Page 12 (http://nationalatlas.gov/elections/elect12.gif).

3. Look up the percent voting for the either the Democrat or Republican candidate in 1996 and 2000. Make a scatterplot and describe the pattern. Are there any unusual clusters of points? Refer to http://nationalatlas.gov/electionsprint.html to aid your discussion. The U.S. Bureau of the Census (http://www.census.gov) produces the Statistical Abstract (http://www.census.gov/statab/www) of the

U.S. The 2002 on-line edition (http://www.census.gov/prod/www/statistical-abstract-02.html), Section 7: Elections, page 6, lists the percent voting for Gore (Democrat) and Bush, Jr. (Republican) in 2000. The 1999 on-line edition (http://www.census.gov/prod/www/statistical-abstract-us.html), Section 8: Elections, page 6, lists the percent voting for Clinton (Democrat) and Dole (Republican) in 1996.

Problem 2.1 (B)

Correlation between measurements on Dogs, Zoo Animals, and Fish

Measurements on several types of animals were recorded in problem 1.2 (D). For one set of dogs, zoo animals, or fish, select two variables. Convert all measurements on each variable to a common scale.

1. Make a scatterplot of your data. Label the axes. Describe in words the pattern that you see.

2. Compute the correlation between the variables. How does the correlation coefficient relate to the graph?

3. If there are any extreme values in either variable, remove that point and recalculate correlation. Does the correlation change much? If so, why? If not, why not?

Problem 2.1 (C)

On-line application for making scatterplots

See the application http://statman.stat.sc.edu/~west/applets/clicktest.html on correlation and scatterplots.

1. Add one point to the graph. No correlation is reported. Add a second point. A correlation of $r = 1$ is reported. Add a third point not in a straight line with the other two. The correlation is less than 1 ($-1 < r < 1$). Write a sentence explaining why $r = 1$ for two points, but $r \neq 1$ for three.

2. Make 3 scatterplots to produce desired correlations.

   (a) Clear the graph. Add 10 points to the graph to create a scatterplot with correlation above 0.90 ($r > 0.90$). If your first attempt is not successful, hit Clear and try again. Examples can be found at http://statman.stat.sc.edu/~west/applets/rplot.html. Print the page to record of your work.

   (b) Clear the graph. Add 10 points to the graph so that you create a scatterplot with correlation between 0.30 and 0.70 ($0.30 < r < 0.70$). Print the page to document your work.

   (c) Clear the graph. Add 10 points to the graph so that you create a scatterplot with correlation between -0.50 and -0.80 ($-0.80 < r < -0.50$). Print the page for your records.

3. See http://www.stat.uiuc.edu/~stat100/java/GCApplet/GCAppletFrame.html. Play the game 5 times. Print the pictures each time to record your scores.

Problem 2.1 (D)

Correlation of prices of sets of stocks at two time periods

A lot of financial information on investment opportunities is on-line. See http://www.fidelity.com. Select "Mutual Funds," then "Learn about Fidelity Funds," and then "Quarterly Performance Reviews."

1. Create a data set. Select 3-4 funds from each of 3-4 categories. Click on your first fund. Go to the bottom of the pop-up window. Record the average annual percent change for 1, 3, 5, and 10 years. Record the same information for the reference indexes, such as the S&P 500 index. Record information for your others funds and any other indexes that appear in comparisons.

2. Plot 3-year percent changes versus 1-year percent changes for your funds. Is the association positive or negative? Are there any outliers? Describe the plot.

3. Compute the correlation between 1- and 3-year percent changes. Relate the correlation to the graph.

4. Repeat the previous two parts for 10- or 5-year percent changes versus 1-year percent changes.

Problem 2.1 (E)

Educational spending by state: scatterplots, correlation, and outliers

Read the story http://lib.stat.cmu.edu/DASL/Stories/EducationalSpending.html concerning the average amount paid to teachers and spent per pupil in U.S. public schools in 1986 for the fifty states plus the District of Columbia in 1986. See http://lib.stat.cmu.edu/DASL/Datafiles/EducationalSpending.html for the data.

1. Describe the overall pattern in the graph and the outlier. What impact does the outlier have on the apparent trend? What are the scales of measurement of the variables? That is, how are they recorded?

2. Will a state law to increase teachers' pay lead to an increase in spending per pupil? Will a state initiative to spend more on students, perhaps through mandating smaller classroom sizes, lead to more pay for teachers? Does the apparent relationship prove causation?

   What else is there? The second paragraph of the DASL discussion concerns regression, analysis of variance (ANOVA), and analysis of covariance (ANCOVA) applied to these data. Linear regression aims to predict one variable as an outcome based on the others. ANOVA and ANCOVA focus on the impact of geographical region on the dollar-valued variables. These are more advanced techniques that would appear at the end of an introductory course or in a second statistics course.

3. Create a current data set comparing characteristics of the educational system by state, make scatterplots, compute correlations, and write a summary of what you find. Are any states unusual? If so, why? Section 4: Education of http://www.census.gov/prod/www/statistical-abstract-02.html, the 2002 on-line edition of the Statistical Abstract of the United States, contains several variables that can be used in scatterplots. Data from 1990 and 2000 on a single variable also could be used. See page 11: educational attainment of the population in 1990 and 2000, page 18: public elementary and secondary enrollment rates in 1990 and 2000, page 21: average public elementary and secondary teacher salaries in 2000, and page 24: per capita public expenditures in 2000. If you classify the states and D.C. into regions, then you can mark points on a hand-drawn or computer-generated plot with symbols or colors corresponding to regions. Maps of the U.S. can be found at http://www.50states.com/us.htm and http://www.lib.utexas.edu/maps (select United States, then a link to a U.S. map down the page).

Problem 2.1 (F)

Scatterplots and classification for three flea beetle species

Can you distinguish flee beetle species based on measurements of their size? In problem 1.3 (C), angle and width were examined separately. The data are at http://lib.stat.cmu.edu/DASL/Datafiles/FleaBeetles.html.

1. Can you distinguish the different species when looking at scatterplots of angle and width? Make a scatterplot with symbols or colored points representing each species.

2. What are the correlations of these measurements overall? What are the correlations in each group? Explain how the differences correspond to the scatterplot.

Problem 2.1 (G)

Readings on correlation and the ecological fallacy

One frequently encounters data available on groups, but questions of interest concerning individuals.

1. See http://stat-www.berkeley.edu/~census/549.pdf for a paper by David Freedman on this subject.

    (a) Define the ecological fallacy.

    (b) Why is it of concern? Describe at least one example.

2. See http://lib.stat.cmu.edu/DASL/Stories/AlcoholandTobacco.html for story and a link to data and http://www.cia.gov/cia/publications/factbook/geos/uk.html for a map.

    (a) Identify the outlier. What impact does it have on correlation?

    (b) Describe how the ecological fallacy applies here.

3. Medical research articles are available on INFOTRAC: search on "ecological fallacy." See "Fluoride and fractures: an ecological fallacy," by Clifford J. Rosen, *The Lancet*, January 22, 2000, v. 355, issue 9200, page 247, or "Socioeconomic deprivation and health and the ecological fallacy," by Ken MacRae, *British Medical Journal*, December 3, 1994, v. 309 n. 6967, pages 1478-1479.

    (a) Describe one example of the ecological fallacy from these articles.

    (b) What groups are involved? What are the variables? Could the data be gathered on individuals?

4. See http://lib.stat.cmu.edu/DASL/Stories/SmokingandCancer.html on smoking and cancer.

    (a) The relationship is strong and quite linear, but does this prove that smoking causes cancer? a

    (b) In this case, how does the ecological fallacy weaken the claim of causation?

2.2   Regression

The table below lists the problem contained in section 2.2.

Problem	Description
2.2 (A)	Prediction of career records by years of sports competition in tennis.
2.2 (B)	Salary versus age for corporate executive officers and small business characteristics.
2.2 (C)	Airline ticket prices versus distance between cities.
2.2 (D)	The impact of a single point on intercept and slope: an on-line demonstration.
2.2 (E)	Estimation of regression lines for randomly generated data and MSE.
2.2 (F)	The impact of subset selection (range restriction) on SAT scores and strength measurements.
2.2 (G)	Regression using summary statistics.

Problem 2.2 (A)

Least squares regression prediction of career records by years of sports competition

The longer a professional sports player is active in a league the more accomplished one would expect the player to be, right? Let's see if it is true for tennis. First, create a data set. From either men's or women's tennis, select a dozen players or more and record several variables.

- Men's Tennis: http://www.atptennis.com/en. Select on "Champion's Race," then a player's name. The the year turned pro subtracted from current year minus one is the number of professional years. Other variables include the number of single's and double's wins, losses, and titles and career prize money.

- Women's Tennis: http://www.wtatour.com. Select "Rankings," then a player's name.

1. Make a scatterplot of a variable versus years pro. Describe the relationship and compute the correlation.

2. Compute the regression equation for predicting the outcome variable in terms of the years pro.

   (a) Interpret the regression coefficient. What does it mean?

   (b) What are the units of measurement on the slope and the intercept? Hint: they are not the same.

3. Predict the outcome variable for the players in your data set. What are the residuals?

4. Make a residual plot of the residuals versus the predictor. Does the regression model fit the data?

5. Repeat the exercise in parts 1-4 above with the predictor variable being the year turned professional.

6. Repeat the exercise in parts 1-4 above with either predictor, but a different outcome variable.

Problem 2.2 (B)

Ages and salaries of CEOs and characteristics of the best small firms

DASL presents salaries and ages of Corporate Executive Officers (CEOs) of the best small firms in 1993. See http://lib.stat.cmu.edu/DASL/Stories/ceo.html and http://lib.stat.cmu.edu/DASL/Datafiles/ceodat.html.

1. Based on the histograms alone can you tell whether the correlation is negative, zero, or positive? Why?

2. Calculate the least-squares regression equation for predicting salary based on age. Exclude the CEO with the missing salary from computations.

3. Write a sentence explaining what the slope coefficient means. What are the units of measurement?

4. What salary is predicted for a CEO with missing salary? What are other 47-year-old CEOs' residuals?

5. Make a residual plot and describe any patterns or lack thereof.

6. Forbes ranks companies (http://www.forbes.com/2002/10/10/200bestland.html). A list of the 200 best small companies is at http://www.forbes.com/static_html/200best/2002/rank.html. Clicking on a company name brings up company information. Use generator pages at http://www.random.org/nform.html or http://www.randomizer.org/form.htm to randomly select 12-20 integers between 1 and 200. Record a pair of variables for companies corresponding to the integers. Pairs of variables include growth in sales (%) and returns on equity (%) at 5 years and at 12 months. Another pair is the recent market price ($) and the P/E ratio (see http://www.investopedia.com/terms/p/price-earningsratio.asp).

   (a) Make a scatterplot of the data.

   (b) Compute the regression line and draw it on the graph.

   (c) Make a residual plot.

   (d) Summarize in a paragraph what you have learned.

Problem 2.2 (C)

Distances between cities and airline ticket prices

Do airlines price tickets according to the distance flown? We can investigate by forming a data set off the Internet and computing regression estimates. Here are three sites that give lowest fares to several destinations.

- Chicago Tribune, http://www.chicagotribune.com, Travel section, Lowest airfares (domestic or international) – a link on the right-hand middle of the page.

- Washington Post, http://www.washingtonpost.com, Travel section, Daily domestic airfares report.

- Los Angeles Times, http://www.latimes.com, Travel Section, Discount Fares under Trip Planning. This site includes information on several points of origin.

Record names and prices for several cities. Find the distances to the destinations (select a departure and destination city at http://www.travelnotes.org/NorthAmerica/distances.htm and press "Show distance").

1. Plot the data. Describe the association between price and distance. Compute the correlation coefficient.

2. Compute the regression equation for predicting price based on distance. What does the slope mean?

3. How much more would you expect to pay to fly an extra 100 miles? 1000 miles?

4. Make a residual plot. Are there any really good deals? Are there any really expensive places to fly?

Problem 2.2 (D)

On-line demonstration of the impact of influential points in regression

See the regression and outlier applet at http://www.stat.sc.edu/~west/javahtml/Regression.html.

1. What is the value of the correlation coefficient? Why is it so high? What is the regression equation?

2. Add a point to the scatterplot so that the regression line has a smaller slope. Report the slope and correlation. Either print the page or draw a picture to report your results.

3. Move the point (click on the screen again) so that the regression line has a higher slope. Report the slope and correlation. Add the new point and line to your picture using a different color or different symbols, such as an x and a dotted line.

4. Can you move the point and increase the correlation coefficient? Describe in words what you do and why the correlation increases.

5. Can you place the point so that the regression slope does not change but the intercept increases? decreases? Describe in words how you do this.

Problem 2.2 (E)

Guess the regression line – an interactive demonstration

See http://www.ruf.rice.edu/~lane/stat_sim/reg_by_eye. In this activity, you place a regression line on the screen by clicking on the screen, holding down the mouse button, moving the mouse pointer, and releasing the button. The program reports the Mean Square Error (MSE), $\sum_{i=1}^{n} e_i^2/(n-2)$, where the e_i's are determined by your line. The minimum MSE $((1-r^2)s_y^2)$ is achieved by the least squares regression line. The spread of the standard errors is described by $\sqrt{\text{MSE}}$.

1. For the first data set, guess the correlation, then press the "show r" button. Were you correct? How could you tell? Describe your experience in a sentence.

2. For the data set, guess the regression line. Guess again. Try to make MSE as smalll as possible. Press the "Draw Regression Line" button. How close were you? Write a sentence about your experience.

3. Repeat the process three more times. Draw rough pictures or print the plots with the lines.

Problem 2.2 (F)

Regression and correlation under restrictions of range: SAT and strength data

See http://www.ruf.rice.edu/~lane/stat_sim/restricted_range/index.html. Many studies are conducted on range-restricted groups. An intuition for relationships between quantitative variables includes an understanding of the impact of subgroup selection. Click the "Begin" box.

1. Choose the SAT data set. Imagine predicting SAT-M based on SAT-V scores. What is the correlation and regression slope? What does the regression slope mean?

2. Move the bar on the left to a verbal score of 500. What happens to the correlation and slope? Move the bar on the left to 675. Write a sentence describing what happens.

3. Select the strength data. Imagine predicting arm strength based on grip. Describe the relationship.

4. Move the bars to 90 and 130. What does this do to the correlation and regression slope?

5. Check the box "Data outside bars." What is the result? Summarizing the strength example in writing.

6. Search on Kobrin at http://www.collegeboard.com/research/home. See College Board Research Note RN-15, "Students with Discrepant High School GPA and SAT I Scores," January, 2002, by Jennifer Kobrin and Glenn Milewski. Why are ranges restricted in these analyses? What is the impact on R^2?

Problem 2.2 (G)

SAT Math and Verbal percentile ranks: regression using summary statistics and Z-scores

The College Board reports mean and standard deviations for Math and Verbal SAT scores and for the total (Math plus Verbal) SAT score based on the entire population of SAT test takers for a given year. See http://www.collegeboard.com/counselors/hs/sat/scorereport/scoredata.html.

1. The variance of a sum of two dependent random variables, V and M, is $\text{var}(V + M) = \text{var}(V) + \text{var}(M) + 2\text{cov}(V, M)$, where $\text{cov}(V, M) = \text{cor}(V, M)\text{SD}(V)\text{SD}(M)$ is the covariance which is equivalent to the correlation times the standard deviations of the two variables. What is the correlation between SAT Verbal and Math scores? That is, locate the three variance terms and solve for the correlation. Remember, the standard deviation is the square root of the variance.

2. Compute the equation of the least-squares regression line for predicting Verbal based on Math. What Verbal scores do you predict for students who score 600, 516, and 400 on the Math exam?

3. Compute the equation of the least-squares regress line for predicting Math based on Verbal. What Math scores do you predict for students who score 600, 504, and 420 on the Verbal exam?

4. Suppose you are predicting variable X based on variable Y. The Z-score for a score of x on variable X is $z_x = (x - \text{mean}(X))/\text{SD}(X)$. The corresponding Z-score for variable Y is $z_y = rz_x$ where r is the correlation coefficient. The corresponding value of Y is $y = \text{mean}(Y) + z_y\text{SD}(Y)$.

(a) If a student scores in the 86th percentile of the Verbal exam, what is her predicted percentile on the Math exam? What is the predicted Verbal percentile for a student at the 86th Math percentile?

(b) What Verbal percentile is predicted for a student at the 43rd Math percentile? What percentile is predicted for a student who scores at the 43rd percentile on the Verbal exam?

(c) Students score 710, 504, and 480 on the Verbal exam. What are their predicted Math scores? Use the percentile method and not the regression equation to make a prediction.

(d) Using percentiles, what Math scores are predicted for students at 710, 516, and 480 Verbal?

2.3 Transformations of Two Quantitative Variables

Problem	Description
2.3 (A)	Linear transformations of temperature and wind speed.
2.3 (B)	Nonlinear transformations of Alaska pipeline calibration data and regression.
2.3 (C)	Wealth of billionaires versus age and sales versus assets for top corporations.
2.3 (D)	A linear and a nonlinear prediction equation for the weight of a horse.

Problem 2.3 (A)

Transformations of temperature and wind measurements

This problem uses current weather data from a site such as http://weather.chicagotribune.com/US/IL.

1. For ten cities, what is the correlation between temperature and wind speed? Make a scatterplot and write a sentence relating the correlation to the plot. Use your state if you like.

2. Change temperature to Celsius (Celsius = (Fahrenheit -32) (5/9)) and wind speed to kilometers per hour (kilometers per hour = 1.6 miles per hour). What is the correlation between temperature and wind speed? Make a scatterplot and write a sentence explaining how it compares to the value in part (1).

3. Change temperature to Kelvin (Kelvin = Celsius +273.15) and wind speed to cubits (ancient Greek) per minute (see http://www.convert-me.com/en/convert/length). What is the correlation on these scales? temperature an wind speed? Explain how it compares to the values in parts (1) and (2).

4. See http://www.wunderground.com/US/NE for weather information in Nebraska (or change the state initials to those for your state). Select 10 cities. What is the correlation between temperature and pressure using the various measuring scales? Make a scatterplot. See http://www.convert-me.com/en/convert/pressure. Inches means inches of mercury.

Problem 2.3 (B)

Alaska pipeline ultrasonic calibration data: transformations in regression

Defects in the Alaska pipeline are measured in the field using ultrasonic measurements, which produce a reading called the depth of the defect in the pipeline. Depth measurements can be verified in the laboratory. See the background (http://www.itl.nist.gov/div898/handbook/pmd/section6/pmd62.htm) at NIST.

1. See the "Fit Initial Model" section. What is the estimated regression line?

2. Does there appear to be any non-linearity in the relationship between laboratory and field measurements? Comment on the scatterplot and residual plots.

3. Look at "Transformations to Improve Fit and Equalize Variances." Write down the equations implied by the transformations. See http://planetmath.org/encyclopedia/Logarithm.html on logarithms.

4. Which transformation seems to produce the plot most appropriate for linear regression? Why?

Problem 2.3 (C)

Nonlinear transformations and regression: billionaires and Fortune 500 companies

This problem considers two DASL sites and lists at *Forbes* related to money. The first concerns age and wealth of Billionaires. The second relates sales to assets for Forbes 500 companies.

1. See http://lib.stat.cmu.edu/DASL/Stories/Billionaires1992.html and read the story.

   (a) For the individuals in one region, make a scatterplot of wealth versus age. Make sure to include units of measurement in your axis labels.

   (b) What is the correlation between wealth and age?

   (c) Make a scatterplot of 1/wealth versus age. Include units of measurement. Does the relationship look more or less linear than before? Why?

   (d) Compute the correlation between 1/wealth and age. Does this make sense for the graph?

   (e) Make graphs and redo calculations for a current sample of billionaires. See http://www.forbes.com/lists/2003/02/26/billionaireland.html for data.

2. See http://lib.stat.cmu.edu/DASL/Stories/Forbes500CompaniesSales.html and read the story.

   (a) For companies from one sector, plot sales versus assets. Include units of measurement.

   (b) Compute the correlation between sales and assets.

   (c) Make a scatterplot of log(sales) versus log(assets). You may use log to the base 10 ($\log_{10}$) or natural log (ln; base e). Does the relationship look more or less linear than before? Why?

   (d) Compute the correlation between log(sales) and log(assets).

   (e) Repeat the exercise with either different variables (sales and employees) or current data from http://www.forbes.com/2002/03/27/forbes500.html.

Problem 2.3 (D)

Horse weight prediction, or how much does your horse weigh?

Few people, including many owners of horses, have scales large enough to determine a horse's weight. A horse's weight, however, is important to know for several purposes, including feeding and administering proper doses of medicine. Associations concerned with horses allow you to estimate the weight of your horse based on a couple of measurements, girth and length, that are easier to make than weight. See section 11.4 for further information on this problem.

1. Read "Estimate your Horse's Weight" at http://www.gaitedhorses.net/Articles/horseweight.html. A linear equation for predicting the weight of a horse based on length and girth is of the form weight $= \beta_0 + \beta_1 \text{length} + \beta_2 \text{girth}$, where β_0, β_1, and β_2 are coefficients.

   (a) What appears to be the coefficient for length? That is, how much does weight (in pounds; lbs) seem to change when length changes by one inch? Calculate the weights for horses with Girth and Length combinations: $(73, 64)$, $(72, 64)$, and $(71, 64)$. Record the weights. Estimate β_1.

   (b) What appears to be the coefficient for girth? That is, how much does weight (lbs) seem to change when girth changes by one inch? Calculate the weights for following horses: $(71, 65)$, $(71, 64)$, and $(71, 63)$. Record the weights. Estimate β_2.

   (c) What appears to be the intercept, β_0? Use the following equation: $\beta_0 = \text{weight} - (\beta_1 \text{length} + \beta_2)$.

   (d) Can this model be used reliably for all sizes of horses? Write a few sentences describing limitations of a simple model like this linear equation for estimating the weights of horses.

2. Read "Weight of a Horse" at http://www.rollinghorse.com/resources/weight.html.

   (a) Calculate the predicted weight of horses with the following Girth and Length combinations: $(73, 64)$, $(71, 64)$, $(71, 63)$. How do the predictions compare to those made previously?

   (b) Write a few sentences describing difficulties that one could encounter when measuring a horse.

3. See http://www.suffolkpunch.com/horses/staindex.html and horses/stalnum.html for measurements on Suffolk Punch stallions. For three of these stallions, compute the predicted weight using both equations. What are the errors in prediction for the two methods?

2.4   Time Series

Problem	Description
2.4 (A)	Time trends in stock market reports: the impact of series length.
2.4 (B)	Trends in global temperature and atmospheric CO_2.
2.4 (C)	Global positioning satellite calibration data.

Additional examples are available under "Time Series" at DASL: http://lib.stat.cmu.edu/DASL/allmethods.html.

Problem 2.4 (A)

Time trends in stock market reports: short- versus long-term

The way in which financial data (prices, values of stock, advertising budgets, wages, amounts sold, etc.) change over time is at least as important as the long-run average or standard deviation. Ignoring the time dimension with such data misses key aspects of the process. See http://finance.yahoo.com. Stocks are indexed by symbols. For example, AAPL and TXN stand for Apple Computers and Texas Instruments. Others include IBM and Dell. Lookup symbols for two or more companies that you think of as being competitors.

1. Look at one of the companies. Choose the 5-day (5d) trend. Describe the pattern over time.

2. Choose the 6-month trend. Is the trend the same as before? If not, how is it different?

3. Enter a competitor's symbol in the comparison box. How do the trends of the two companies compare?

4. Look at the five year trend, or however long is possible. What is apparent in the longer trend that was not visible over a shorter duration?

Problem 2.4 (B)

Time series: seasonal and long-term patterns in temperature and CO_2

There has been concern expressed by scientists, government officials, and the media over the emission of greenhouse gases and the warming of the Earth's climate. What evidence is there? Any evidence is going to be quantified as time series measurements. Describe trends and patterns from the sources below.

1. See the links to "Climate, what is the problem?" at the Environmental Protection Agency (EPA) (http://yosemite.epa.gov/oar/globalwarming.nsf/content/index.html), and the "glacial cycle" pages at the National Oceanographic and Atmospheric Administration (NOAA) Paleoclimatology Program (http://www.ngdc.noaa.gov/paleo/primer_history.html). Is the temperature rising?

2. Is the level of carbon dioxide (CO2) in the atmosphere rising? See the first graph at The Exploratorium (http://www.exploratorium.edu/climate/atmosphere/index.html) and the CO_2 Monthly Means (JPEG) at the National Oceanic and Atmospheric Administration (NOAA), Climate Monitoring and Diagnostics Laboratory (CMDL) (http://www.cmdl.noaa.gov/ccg/figures/figures.html) Comment on seasonal cycles as well as the trend in the average. See also *In Brief – The U.S. Greenhouse Gas Inventory* by the EPA: http://yosemite.epa.gov/oar/globalwarming.nsf/content/emissions.html.

3. Local weather in the short term (minutes, hours, days, weeks, months) also exhibits trends and cycles. What time-related information is at http://www.wunderground.com? Select a city and then an individual weather station.

Problem 2.4 (C)

Dependent multivariate time series: Global Positioning Satellite calibration data

It is possible now to locate positions on the Earth based on signals from global positioning system (GPS) satellites orbiting the planet. Since the satellites have their own standards for keeping time, they are monitored continuously and compared to the time kept at NIST: http://www.boulder.nist.gov/timefreq/service/gpstrace.htm. The data from comparisons of several GPS satellites are made available daily. Press the "Get Data" button on this page. Find the table pertaining to individual GPS satellites. The mean time offset, measured in nanoseconds (one-billionth of a second), is average of the discrepancies in 10-minute time averages for a day.

1. Look at the picture and describe the pattern.

2. Click on "View" beside the line in the table below for an individual satellite. Compare the pattern for several satellites. Are there any gaps due to data not being transmitted or received?

3. Return to the first page at this site. Select "20 days back" and look at the time series. Describe the pattern overall. How does it compare to the 1-day series?

4. Click on "View" for an individual satellite. How do patterns compare across satellites?

5. Repeat the last two parts for "200 days back."

Chapter 3

Categorical Data

Introduction

Measurements on categorical data and data with a few possible outcomes, such as the six values on a standard six-sided die, can summarized by the proportion or number of times each category or value is observed. If two categorical variables are measured on the same set of units or individuals, then two-way cross classifications of the data yield two-way tables of counts. Each count represents the number of cases reporting a certain combination of responses on the two variables. The proportion of the cases classified into a particular cell or set of cells in the table (a joint relative frequency) or the proportion of the cases in a certain category of one variable within a row or column defined by the other variable (a conditional relative frequency) also can be reported. Bar graphs and segmented bar graphs can be used to display these data.

Remarks

After describing the relationship between the responses to two variables, try dividing the cases according to a third variable and analyzing the relationship separately by group. Results can change in dramatic ways. Simpson's paradox is the name given to the phenomenon observed when heterogeneous groups are combined and apparent relationships are changed or even reversed. Most problems concerning categorical data can be approached using proportions, fractions, or percents or using tables of counts. Rephrasing a problem in different terms sometimes is helpful in finding its solution.

Outline

Section 3.1 presents problems concerned with describing measurements on one, two, or more categorical variables. Section 3.2 provides examples of and questions about Simpson's paradox. Section 3.3 contains examples of maps, which are a graphical tool not covered in the other sections and can be used to display categorical data with a geographical association. Pie charts, which can be used to represent the proportion of observations in a category, are presented in chapter 5 on probability.

3.1 Describing Categorical Variables

The following table outlines the problems contained in section 3.1.

Problem	Description
3.1 (A)	Comparing counts, percentages, and proportions using BLS wage and salary data.
3.1 (B)	Conditional relative frequencies and tables of BLS demographic and work status.
3.1 (C)	Number and rates of cases of Lyme disease in the United States.
3.1 (D)	Aggregation of small areas and proportions and rates of HME in Iowa deer.

Problem 3.1 (A)

Bureau of Labor Statistics Wage and Salary data by State: Counts, Percentages, and Proportions

The Bureau of Labor Statistics (http://www.bls.gov) is the official source of wage data for the U.S.: http://www.bls.gov/bls/blswage.htm. Data are reported by state: http://www.bls.gov/oes/2001/oessrcst.htm.

1. Select your state. How many people are employed? Pick three occupations. How many people are employed in these three occupations? What percentage of those employed in your state work in these occupations? Express your answer both as a percentage and as a fraction or decimal.

2. Choose a state that you think will have similar distributions of occupations as your state. See if you are correct. That is, compute the proportions in the three occupations in your comparison state.

3. Choose a state that you think will have very different proportions in the three occupations. Calculate the proportions in the second comparison state.

4. Make a three-by-three table of counts with rows defined by occupations and columns defined by states. Enter the counts into the table. Produce a second table that contains the percentage of those employed by occupation and state in each cell of the table. Make the two tables below and write a sentence contrasting the occupations or states for each.

   (a) Produce a third table that reports the percentage in each occupation by state.

   (b) Produce a fourth table that reports the percentage in each state by occupation.

Problem 3.1 (B)

Bureau of Labor Statistics Wage and Salary data: conditional relative frequencies and tables

The Bureau of Labor Statistics (BLS) also reports demographic characteristics of the labor force based on the Current Population Survey (http://www.bls.gov/cps/home.htm). The tables are located at http://www.bls.gov/cps/home.htm#tables. The numbers in the tables are the numbers of thousands of people.

1. See http://www.bls.gov/cps/home.htm#charemp for characteristics of employed people in the U.S.

   (a) What percent of the workforce is female? male?

   (b) What percent of the workforce is 16-20 years old? over 20?

   (c) Which group, female or male, has a larger percentage of its workforce members under 20 years old? over 20? Make a table with female and male on the rows and non-overlapping age groups above the columns and fill-in the cells of the table.

2. See http://www.bls.gov/cps/home.htm#charunem for characteristics of the unemployed in the U.S. Unemployed is different from not in the labor force; see http://www.bls.gov/cps/cps_faq.htm.

   (a) For 2001, make a table with Black, White, and Other on the rows and non-overlapping age groups on the columns. Fill in the counts based on the BLS data.

       i. If you randomly selected an unemployed person in the U.S., what is the chance the person would be black? white? neither black nor white?

       ii. Which group (black, white, or other) has the highest percentage of its unemployed in the youngest age group? in the oldest age group?

       iii. Which age group has the highest percentage of its unemployed persons among the population that is neither black nor white?

   (b) See http://www.bls.gov/cps/#pnilf concerning those individuals not in the work force. Make a table with "Do not want a job now" and "Want a job" on the rows and non-overlapping age groups on the columns. Describe this table using percentages and proportions.

   (c) See http://www.bls.gov/cps/#multjob for information on individuals working multiple jobs. Make a table cross classifying subjects by non-overlapping age groups and by sex in 2001. Describe this table using percentages and proportions.

Problem 3.1 (C)

Reported Cases of Lyme Disease: proportions and rates versus totals

The American Lyme Disease Foundation (ALDF; http://www.aldf.com) and the Center for Disease Control (CDC; http://www.cdc.gov/ncidod/dvbid/lyme/index.htm) provide information about Lyme Disease. The ALDF provides a map and table (http://www.aldf.com/usmap.asp) of reported cases of Lyme Disease.

1. What fraction of cases in 1997 were reported in Connecticut, in Vermont, in New York, and in Texas? How about in 2000?

2. Of the reported cases in Connecticut, what percent were reported in 1997? 2000?

3. What is the chance that a randomly selected case from North Carolina, South Carolina, or Tennessee was reported in 1997? in 2000?

4. Vermont has a smaller number of cases in 2000 than does New York, but New York has a lower incidence rate. Why is this? For state populations, see the U.S. Bureau of the Census' QuickFacts site: http://quickfacts.census.gov/qfd.

5. Create a table with one column for each year and four rows, one for each category 0, 1-5, 6-25, and more than 25. Tabulate the number of states in each of the cells of the table.

   (a) What percent of states reported no cases of Lyme disease in 1997? in 2000?

   (b) What fraction of states reported more than 5 cases of Lyme disease in 1997? in 2000?

   (c) How is the distribution of the number of cases (in these categories) by state changing over time?

Problem 3.1 (D)

Iowa HME deer study: proportions overall and in smaller areas

Human monocytic ehrlichiosis (HME) is a disease that was recently identified and is carried by the Lone Star tick. Deer blood can be tested for the presence of bacteria that carry HME. A study at the University of Iowa tested deer blood in 1994 and 1996. See http://www.uhl.uiowa.edu/HealthIssues/HME/intro.html and http://www.uhl.uiowa.edu/HealthIssues/HME/index.html for information on the study. See http://www.ipm.iastate.edu/ipm/iiin/tlonesta.html and http://www.ent.iastate.edu/imagegal/ticks/aamer for information on the Lone Star tick.

1. For the whole state of Iowa, what percent of samples that were tested were tested in 1994? in 1996?

2. Make a table for the whole state of Iowa that has 1996/1994 on the left side and Positive/Negative on the top. What percent of cases were Negative and tested in 1994?

3. What fraction of samples were positive in 1994? 1996? What fraction were negative in these years?

4. What is the chance that a randomly selected positive case was tested in 1994? 1996?

5. Given that a randomly selected case is negative, what is the chance that it was tested in 1994? 1996?

6. Find a county that had a higher positive rate in 1994 than in 1996. Find a county that had a higher positive rate in 1996 than in 1994. Report the table and percentages for the county.

7. Find a county that had more than fifty percent of its positive tests in 1994. Find a county that had more than fifty percent of its positive tests in 1996.

8. Some of the counties had few tests performed in 1994 and 1996. As a result, the percentages can change a lot between time periods even though not many deer test positive. One option is to combine data from neighboring counties in order to produce a composite estimate. The Northwest corner of the states contains several counties with small sample sizes. Create a table for these counties combined. You can choose whether you want to combine 9, 12, 16, or 19 counties. Comment on how the results when combined compare to the rates in the individual counties.

3.2 Simpson's Paradox

Simpson's Paradox is named after E.H. Simpson who published an article in the Journal of the Royal Statistical Society in 1951. A search of the archives of the AP Statistics Newsgroup (see http://mathforum.org/discussions/epi-search/apstat-1.html) using the keyword "simpson" (lowercase) produces a few items. The letter by Michael Larsen on Monday, May 4, 1998, briefly describes Simpson's article. The main idea of the paradox is that combining data from heterogeneous groups can produce composite data that exhibit different relationships between variables than in all of the separate groups. Simpson used data on cards (categorical data) to illustrate the phenomenon, but the reversal or alteration of relationships caused by the aggregation of data generally should be anticipated. The table below lists the problems in section 3.2.

Problem	Description
3.2 (A)	Rates of success for two players in two halves of a game.
3.2 (B)	Rates of survival at two hospitals for two groups of patients.
3.2 (C)	Journal of Statistics Education examples of Simpson's paradox in South Africa.
3.2 (D)	Berkeley graduate admissions, "Ask Marilyn," and other examples.
3.2 (E)	A generalization applied to evaluations of teachers, investment management.
3.2 (F)	Articles on smoking and mortality and on diabetes and insulin treatments.

Problem 3.2 (A)

A Simpson's paradox using success rates for two players in two halves

An example of Simpson's paradox is provided by http://www.cate.org/sms99/writ99/simprdx.htm.

1. Explain why Simpson's paradox applies to this situation.

2. Imagine that player B had made 40 attempts in the second half with 16 successes. What would be the combined total and rate for player B?

Problem 3.2 (B)

Simpson's Paradox and categorical data on hospital survivorship

A medical example of Simpson's paradox is provided by http://www.ma.iup.edu/~zhang/simpson.html.

1. Explain why Simpson's paradox applies to this situation. The questions on the site provide a guide.

2. Increase the number in the good condition in Hospital B by a factor of 10: replace 30 by 300 and 70 by 700. Do we still observe Simpson's paradox in effect? Explain what has changed and why it matters.

   Further activity: This site links to a calculator: http://www.ma.iup.edu/~zhang/psimpson.html.

Problem 3.2 (C)

Simpson's paradox and data on children in South Africa

The *Journal of Statistics Education* reports an example of Simpson's paradox in a study in South Africa. See http://www.amstat.org/publications/jse/secure/v7n3/datasets.morrell.cfm.

1. Read this table. Copy the tables and demonstrate that Simpson's paradox is in effect here.

2. Explain in this example why the associations reverse when the data are aggregated.

Problem 3.2 (D)

Simpson's paradox, Berkeley Graduate admissions, and a question to Marilyn

In the cases below, write a sentence or two to document the occurrence of Simpson's phenomenon.

1. University of California, Berkeley, Graduate admissions (with questions): http://www.math.uah.edu/siegrist/ma487/simpson.html

2. Discussion of an article by Marilyn vos Savant, "Ask Marilyn," April 28, 1996, *Parade Magazine*, page 6: http://cq-pan.cqu.edu.au/schools/smad/simpadox.html See also http://www.dartmouth.edu/~chance/chance_news, search for "Simpson's," select "5.07 Ask Marilyn."

3. See examples at the Exploring Data web site: http://exploringdata.cqu.edu.au/sim_par.htm.

Problem 3.2 (E)

Generalizations of Simpson's paradox: evaluations of teachers and Investment Management

Generalizations of Simpson's paradox can involve multiple groups and categorical variables with multiple levels. Here are two examples. In each case, document the occurrence of the phenomenon and write a sentence or two explaining clearly why it happens in each example.

1. An example on student ratings and mathematics ability (with questions): http://www.oswego.edu/~srp/stats/simpsons_wk_1.htm.

2. A paper on Investment Management http://www.uic.edu/~gib/simp.pdf.

Problem 3.2 (F)

Articles involving Simpson's paradox and categorical data

Simpson's paradox affects data analysis and study conclusions in research articles. In INFOTRAC, search on "Simpson's paradox."

1. One example involves data on smoking: "Ignoring a covariate: an example of Simpson's paradox," by David R. Appleton, Joyce M. French, Mark P.J. Vanderpump. *The American Statistician*, November, 1996, v. 50 n. 4, pages 340-1. Compute the 20-year survival rates for smokers and nonsmokers overall. Compute the 20-year survival rates for smokers and nonsmokers within age groups. What do you conclude? What impact does a selection effect have on older women?

2. A second example involves insulin treatments and diabetes: "Confounding and Simpson's paradox," by Steven A. Julious and Mark A. Mullee. *British Medical Journal*, December 3, 1994, v. 309, n. 6967, pages 1480-1482. Compare survival rates for the non-insulin dependent and insulin dependent patients overall. Compare survival rates among the patients 40-years-old or younger and the patients over 40 years of age. Note: this article used some advanced statistical methods, which would be familiar to doctors and researchers trained in medical/biostatistics.

3.3   General Graphical Methods

There are numerous ways to present qualitative and quantitative data graphically. It is beyond the scope of this work to even begin to introduce the range of options. It is worth noting, however, that not all graphics accurately communicate what they claim to portray. Other graphics, although visually interesting, fail to portray important dimensions of some problems. Problem 3.3 (A) provides several examples. The other problems concern what can and cannot be seen in maps. Problem 2.1 (A) included links to maps concerned with voting. The table below describes the problems in section 3.3.

Problem	Description
3.3 (A)	Good and bad statistical graphics.
3.3 (B)	NASS maps of agricultural production in the U.S.
3.3 (C)	Department of Interior West Nile Virus maps.
3.3 (D)	CDC human health maps: per capita rates, population density, and smoothing.
3.3 (E)	State versus county unemployment rates at BLS.

Problem 3.3 (A)

Good and bad statistical graphics

A web site dedicated to the graphical presentation of statistical information is the *Gallery of Data Visualization: The Best and Worst of Statistical Graphics*, by Michael Friendly of York University: http://hotspur.psych.yorku.ca/SCS/Gallery.

1. Select two bad graphical displays and give two reasons each why they are inaccurate or misleading. How could they be presented in order to be clear?

2. Select two good graphical displays and give two reasons each why they aid communication.

Problem 3.3 (B)

Agricultural maps at the National Agricultural Statistics Service: acres and bushels per acre

Maps at U.S. Department of Agriculture, National Agricultural Statistics Service (NASS), show crop acreage and yield by county in the United States. See http://www.usda.gov/nass/aggraphs/cropmap.htm.

1. Select one crop. Where is it grown?

2. For your crop, where is the area of greatest productivity, or bushels per acre?

3. How does the area where it is grown and productivity relate to precipitation and temperature? See http://www.epa.gov/ceisweb1/ceishome/atlas. Look at the precipitation atlas: select "Land," then "Landscape types," then "Average annual precipitation." The map depicting average annual temperature is linked to the same page as the precipitation map.

4. How would you estimate total production of the crop in the U.S. based on these two maps?

Problem 3.3 (C)

Department of Interior West Nile virus maps of qualitative and quantitative data

Positive identifications of the West Nile virus have been made in several states and counties in the United States. See the following sites at the U.S. Geological Survey, Center for Integration of Natural Disaster Information: http://cindi.usgs.gov/hazard/event/west_nile/west_nile.html.

1. Look at the map for birds. According to this map, what fraction (approximately) of the counties in the U.S. have had birds test positive for West Nile virus? Is the true proportion likely to be lower, higher, or about the same?

2. Are all the counties the same? That is, what would you want to know about the counties to distinguish them in terms of West Nile virus from one another?

3. How does time figure into this graph, if at all? How could you incorporate time?

4. Look at the human, bird, and mosquito maps. There seem to be some human cases in Texas counties without bird or mosquito cases. How is that possible? Do you think reporting standards and data gathering procedures are the same for these three groups?

Problem 3.3 (D)

Smoothing of CDC maps of per capita rates and population density

The Center for Disease Control provides atlases related to mortality for the United States: http://www.cdc.gov/nchs/products/pubs/pubd/other/atlas/atlas.htm. Look at the links at the bottom of the page. Data on maps are reported for white males, black males, white females, and black females. Data are available for other groups, but sample sizes presumably are quite small except in certain metropolitan areas. A population density map is available from the U.S. Bureau of the Census: http://eire.census.gov/popest/gallery/maps/co_01_04.php.

1. Where is the population the most dense in the United States?

2. Why is it important to express the occurrence of disease as a rate per so many people?

3. Why might it be of interest to know the total number of cases

4. "Smoothing" is a statistical technique for combining results over geographic areas and groups that are similar in order to produce more stable estimates. Many procedures exist for smoothing data such as those used to produce the health maps. Describe the impact of smoothing on maps for one disease. The cross-hatched areas on the maps are areas in which sample sizes are small.

Problem 3.3 (E)

Bureau of Labor Statistics unemployment maps: county versus state level data

The Bureau of Labor Statistics (http://www/bls.gov) provides maps depicting unemployment rates at the state and at the county level. See http://www.bls.gov/opub/mapbook/home.htm and look at the two charts.

1. Which states experienced the greatest reductions in unemployment in the time period 1990-1999? Which states or regions showed the least change? How did your state perform?

2. Identify some counties for which unemployment dropped significantly but the state rates increased or did not see a great change.

3. Identify some counties that remained nearly constant in states for which unemployment either dropped significantly or increased drastically.

4. Write a sentence describing the value and limitation of both maps. Remember that the rates are based on survey information and sample sizes are larger at state levels than at county levels, but counties can be quite different from one another within a state.

Chapter 4

Research Designs

Introduction

Researchers gather data and conduct studies in order to learn about the world and the beings, human and otherwise, in it. In order to learn as much as possible and to produce results that will be taken seriously, researchers must carefully design their studies. A well-designed study increases the chance of finding something interesting, the opportunity for exploring alternatives, and the likelihood that important questions are answered conclusively. The highest standard for establishing causation is the randomized-controlled double-blind experiment used fields such as medicine and agriculture. Random assignment of subjects to treatment groups avoids selection bias. Double blinding avoids other biases on the part of, in a medical study, the patients, the researchers, and the evaluating doctors. Blocking or matching can increase the precision of overall results by forcing treatment groups to be similar or equal on key variables.

Surveys are undertaken for the purpose of describing a population. Governments, large corporations, news agencies, interest groups, and individuals conduct numerous surveys every day. Surveys can be taken of almost anything: people, groups, plots of land, samples of air and water, and products. Surveys that collect data from randomly selected members of well-defined and comprehensively listed populations are of the highest quality. Many surveys, however, do not control the random selection process and encounter selection bias. In some cases, the population under study is vaguely defined or un-listable.

Studies that are conducted for the purpose of proving that something causes something else to happen, but are not able to use random assignment to conditions as in experiments, are referred to here as observational studies. The data used in observational studies are often collected in surveys, but are not always the result of random selection. Observational studies are prone to confounding bias. Care must be taken in their design and interpretation if there is any hope for making valid causal conclusions. Blocking or matching for control can be incorporated into the design. Variables can be collected for use in statistical adjustment for uncontrolled group differences. A great deal of observational data are readily available and can be used to gain insight into challenging problems.

Remarks

Some terminology is need in order to describe an experiment. Groups of subjects, to which different treatments have been applied, are being compared. The concept of confounding and its importance is critical. Confounding involves both a difference of the confounding factor across treatment groups and an impact on an outcome. Randomization can be implemented in various ways and eliminates bias due to confounding. A further important design choice is blocking or matching to decrease variance. Simulation can be used to understand the control of variability blocking. Simulation allows the examination of repeated hypothetical randomizations. All examples and discussion of design should be related to the context of an applied research question concerned with a outcome of interest.

A survey is used to collect a sample for the purpose of describing a population. Both the sample (the group that actually can be described) and the population (the larger group that we want to study) are defined in research studies. There are several types of bias, including nonresponse, response and coverage bias. Of particular importance is selection bias. If a researcher does not randomly select the sample and interview the selected sample, then there likely is selection bias: the sample will be systematically different from a truly random selection of the population. A factor causing bias (any type) affects the characteristic being measured ("the outcome") and has different levels or distributions in the interviewed and noninterviewed groups. The purpose and precise implementation of random selection should be described in an actual survey context. Further design choices are important in practice. Stratification reduces variance, but does not remove bias. Cluster sampling is often convenient and does not cause bias, but does increase the variance of estimators. Simulation can be used to illustrate the advantages of probability over nonprobability sampling and to contrast various probability sampling designs.

Observational studies are comparative studies in which randomization of treatment assignments is not possible. Even if a variable is strongly associated with an outcome, one cannot automatically attribute cause to it; association does not equal or prove causation. A clear concept of confounding is fundamental to understanding the difficulties involved in making conclusions from observational data. Not every difference between "treatment" or observational groups is confounding. Observational studies are useful when they control group differences with blocking or matching. Examples of Simpson's paradox in section 3.2 illustrate the advantage of blocking to remove confounding bias. Some observational studies focus on the same populations and variables as experiments and surveys. There is a need, therefore, to distinguish between studies that merely demonstrate associations and those that prove causation.

Outline

Section 4.1 presents problems concerning terminology and concepts of experiments. Section 4.2 focuses on sample surveys, both good and bad. Section 4.3 contains discussion of and problems about observational studies. Further examples of studies are encountered throughout the book. The Companion Website can be accessed at http://larsen.duxbury.com. Please enter the Serial Number from the back cover when prompted.

4.1 Experiments

Problems in section 4.1 are listed in the table below.

Problem	Description
4.1 (A)	Parkinson's disease studies.
4.1 (B)	Aromas and test taking performance.
4.1 (C)	Garlic bread and family harmony.
4.1 (D)	Kicking footballs and helium.

Problem 4.1 (A)

Clinical trials and Parkinson's disease

The Parkinson's Disease Foundation (http://www.pdf.org) describes the symptoms and treatments of Parkinson's disease. Several articles on clinical trials can be found in INFOTRAC.

1. Search on "Multidisciplinary rehabilitation for people with Parkinson's disease: a randomised controlled study," by D.T. Wade *et al.*, *Journal of Neurology, Neurosurgery and Psychiatry*, February, 2003, v. 74, issue 2, pages 158-162.

   (a) Define the treatment groups and treatments.

   (b) What outcomes are measured?

   (c) How is randomization used? In your own words, describe why randomization is important in this example. Is there any evidence that randomization was effective?

   (d) What are some other variables recorded on the patients?

2. Select "Sham surgery controls: intracerebral grafting of fetal tissue for Parkinson's disease and proposed criteria for use of sham surgery controls," by R.L. Albin, *Journal of Medical Ethics*, October, 2002, v. 28, issue 5, pages 322-325.

   (a) What are the treatments and outcomes?

   (b) What are some arguments against using sham surgeries in research?

   (c) What are some arguments in favor of the use of sham surgeries?

3. Search on "Parkinson's disease and nurses" and select "Effects of community based nurses specialising in Parkinson's disease on health outcome and costs: randomised controlled trial," by Brian Jarman *et al.*, *British Medical Journal*, May 4, 2002, v. 324, issue 7345, pages 1072-1075.

   (a) What are the treatment groups and outcomes?

   (b) What are some variables that could be used to set up blocks of patients? Describe a study design, including the randomization, using blocking.

Problem 4.1 (B)

The impact of aromas on learning and test taking

DASL contains a story (http://lib.stat.cmu.edu/DASL/Datafiles/Scents.html) concerning the the impact of smelling a floral scent on learning and test taking. In the data file, U-Trials are unscented, and S-Trials are scented; see http://lib.stat.cmu.edu/DASL/Datafiles/Scents.html. The first set of three times is for the U-Trials, and the second set is for the S-Trials.

1. Describe the design. What are the units on which the study is done? What are the treatments?

2. How is matching used? Where is the randomization applied?

3. Does the randomization fulfill its promise? That is, are the distributions of covariates 'balanced' between the groups receiving the treatments in the two orders Compute the average age, the percent female, and the percent nonsmoker for the individuals who received unscented trials first and second.

4. The key to seeing the advantage of matching numerically is to compare results on unscented and on scented trials for the same person. Take the third trial of each type. Which is bigger, the differences in average of these two final trials among different people or the change in time for individuals? Display the association between the final trials by making a scatterplot and computing a correlation. Write a paragraph describing the use of matching and randomization in this study.

5. An important issue in evaluating a study is its relation to real situations. Describe briefly a year-long study to examine the impact of aromas on learning.

Problem 4.1 (C)

The impact of garlic bread aromas on family harmony

Read the brief report by the Smell and Taste Treatment and Research Foundation of Chicago on the impact of garlic bread at dinner and positive family interactions: http://www.smellandtaste.org/garlic.htm.

1. Describe a study design using matching. What are the units in the study? What are the treatments? Describe a possible advantage of matching in this study.

2. How could randomization be used? Why would it need to be used?

3. What covariates or independent variables could be measured? You should be able to list more than two examples. Include in your description how these variables would be recorded.

4. It is difficult to conduct research on people and many animals, because they often know an attempt is being made to influence them. They also are aware that they are being measured or observed, and might behave abnormally. Do you think this is an issue in the garlic bread study? If so, can you suggest an alternative study design? Suppose that your study can last more than two dinners and that data can be collected in ways other than by an observer.

Problem 4.1 (D)

Helium versus air in footballs

DASL reports on a small experiment (http://lib.stat.cmu.edu/DASL/Stories/Heliumfootball.html) of punting footballs when one football is filled with air and the other with helium. A punt is a type of kick that begins with the football player (the punter) holding the football and dropping it onto his or her foot. In a drop kick, the ball is bounced off the ground before it is kicked. In a kick-off, the player kicks the ball off a tee on the ground which supports the ball in an upright position. Helium is lighter than air. The study used one punter who alternated between the helium- and the air-filled footballs. The 39 trials of the two balls were conducted in one session. The data can be found at http://lib.stat.cmu.edu/DASL/Datafiles/Heliumfootball.html.

1. What are the treatments? How does this study use matching? How could randomization be used in the context of this design?

2. Suppose four football players who have been trained to punt the football are available for your study. Describe a research design that utilizes the four players. Suppose you will conduct a total of 40 trials with each football. Why might you want to use 4 kickers instead of 1? Would you allow practice kicks?

3. Suppose four football players who have been trained to punt, drop-kick, and kick-off are available for your study. There are two treatment factors: helium versus air and type of kick. Describe a research design that allows each player to make 18 kicks. How could you use a 6-sided die to randomize the order of different treatment combinations?

4. Suppose two football players trained in the three kicking styles are available at each of five colleges or universities. What are all the factors involved in this study? Can you suggest any other factors that you might like to control or record? Design an experiment involving as many factors as you can.

4.2   Surveys, including Opinion Polls

Discussion of how to conduct good surveys can be found at the Survey Research Methods Section of the American Statistical Association (http://www.amstat.org/sections/srms) and Public Agenda Online (http://www.publicagenda.org/aboutpubopinion/aboutpubop.htm). Problems in section 4.2 are listed below.

Problem	Description
4.2 (A)	Gallup Poll.
4.2 (B)	Internet and forestry surveys.
4.2 (C)	Current Population Survey.
4.2 (D)	Survey of Doctoral Recipients at NORC.
4.2 (E)	Census of Agriculture.
4.2 (F)	A survey of AP teachers.

Problem 4.2 (A)

The Gallup Poll

The Gallup Poll (http://www.gallup.com) is one of the most recognized surveys.

1. See "How are polls conducted?" at http://www.gallup.com/help/FAQs/poll1.asp.

   (a) What is the fundamental goal of a survey?

   (b) Major and minor league baseball teams exist in every state in the U.S. The season lasts several months. Would interviews conducted in every state over a period of a few days be representative of all people in the U.S.? Would increasing the sample size and duration of sampling produce a representative set of interviews?

   (c) What population is covered by the Gallup Poll? What segments of the population are excluded?

   (d) What is random digit dialing?

   (e) The standard error (SE) of a sample proportion is approximately $\sqrt{p(1-p)/n}$, where n is the sample size and p is the actual proportion. For $p = 0.5$, what are the SEs for $n = 1000$ and 4000?

   (f) Describe a surveying task other than selecting the respondents.

2. Gallup interviewed in Islamic countries in 2002. See http://www.gallup.com/poll/summits/islam.asp and http://www.gallup.com/help/FAQs/answer.asp?ID=28.

   (a) How were the populations of each country stratified?

   (b) What were the primary sampling units, or the first areas randomly selected?

   (c) What other levels of sampling were conducted?

   See the Council of American Survey Research Organizations (http://www.casro.org) and the American Association of Public Opinion Research (http://www.aapor.org) for links to other organizations.

Problem 4.2 (B)

Internet and volunteer surveys: examples of nonprobabilty sampling

Numerous Internet sites and the U.S. Forest Service collect data from the people who view or visit their sites.

1. Look at TV network web pages for a "poll" or opportunity to vote. For example, http://www.abc.com (Watercoooler Poll), http://www.cbs.com (Star Search), http://www.fox.com, or http://www.nbc.com (http://www.nbc.com/nbc/Days_of_our_Lives/poll.shtml).

   (a) Define a population that this sample could represent. Is the sample representative of the population? Why or why not?

   (b) Describe a survey (telephone, mail, or other) that would be more representative of the population than the Internet data collection at the networks.

   (c) What would be one advantage for these companies of maintaining an on-going Internet poll?

2. The National Parks Conservation Association (http://www.npca.org/take_action/poll/menu.asp, NPCA) conducts on-line polls on an ongoing basis. Select one poll topic.

   (a) Define a population you would like to study. Does the NPCA poll provide an unbiased representation of this population? Why or why not?

   (b) In what sense are the questions leading the respondents toward an answer?

   (c) Describe an alternative study design that would produce a more representative sample.

3. The National Resources Defense Council (http://www.nrdc.org/media/pressreleases/000803.asp) conducted a survey with the NPCA and the Yosemite Restoration Trust.

   (a) What questions seem to have been asked in this survey?

   (b) Is much of the population in the state of California affected by traffic around the Merced river? Do you think that would affect the survey results?

   (c) Describe two survey design details you would like to know to evaluate the quality of this survey.

4. The USDA Forest Service gathers information in various ways. Comment on the degree to which the methods below provide representative information on well-defined populations. Comment also on the value of the various forms of data gathering.

   (a) Customer service comments: http://www.srs.fs.usda.gov/customer.

(b) Voluntary use survey:
   http://www.fs.fed.us/r9/hoosier/news_releases/general_releases/visitor_use_survey%20092702.htm

(c) National Use Monitoring survey:
   http://www.fs.fed.us/recreation/programs/nvum/reports/year2/R1_F16_lolo_reportf.htm

Problem 4.2 (C)

The Current Population Survey

Most U.S. economic indicators are based on the Current Population Survey (CPS; http://www.bls.gov/cps).

1. What population is covered by the CPS? See http://www.bls.gov/cps/cps_over.htm#coverage What are some uses of the data? See http://www.bls.gov/cps/cps_over.htm#uses.

2. For the next four parts, see http://www.bls.gov/opub/hom/homch1_e.htm. How many primary sampling units (PSUs) are selected? How are the PSUs defined?

3. What characteristics define the PSU strata?

4. How many ultimate sampling units (households) are selected?

5. How and why is the sample rotated? See also http://www.bls.gov/opub/hom/homch1_f.htm.

6. Discuss two limitations of the survey. See http://www.bls.gov/opub/hom/homch1_j.htm.

Problem 4.2 (D)

Survey of Doctoral Recipients at NORC

NORC (http://www.norc.org), formerly the National Opinion Research Center at The University of Chicago, is a nonprofit organization and conducts the Survey of Doctorate Recipients (SDR) for the National Science Foundation (NSF). The SDR is one of three surveys in the NSF's SESTAT system (http://sestat.nsf.gov). See http://srsstats.sbe.nsf.gov/docs/techinfo.html for some design details.

1. What population is covered by the SDR? What is the sampling frame or list?

2. What fraction of the population was sampled in 1993? What was the response rate?

3. What were some questions asked? See http://srsstats.sbe.nsf.gov/docs/source.html#instruments.

4. What is hot deck imputation? See the technical information.

   NORC's other studies include "America Rebounds: A National Study of Public Response to the September 11th Terrorist Attacks" and the Florida Ballots project (http://www.norc.org/fl/index.asp).

Problem 4.2 (E)

The Census of Agriculture and the Farm Identification Survey

The National Agricultural Statistics Service conducts an agricultural census every 5 years.

1. See the frequently asked questions at http://www.nass.usda.gov/census. What is a farm?

2. See http://www.nass.usda.gov/census/census02/preliminary/fisindex.htm Does the farm identification survey sample from a population, or is the term survey being used colloquially?

3. For what geographical units are results reported? See highlights, congressional districts, and rankings.

4. What is an area sampling frame? See http://www.nass.usda.gov/research/AFS.htm.

5. How are strata defined for some states? See http://www.nass.usda.gov/research/stratafront2b.htm.

Problem 4.2 (F)

A survey of AP teachers

The College Board Research Report 2002-10, *What Are the Characteristics of AP Teachers? An Examination of Survey Research*, (http://www.collegeboard.com/repository/200210_20717.pdf) reports on responses from 32,109 teachers.

1. What population is the College Board studying?

2. Is there a list for this population? How does the College Board proceed?

3. Is this a random selection of AP teachers? Discuss limitations of this study. Do you have any suggestions for overcoming the limitations?

4.3   Observational Studies

Problems in section 4.3 are listed in the table below.

Problem	Description
4.3 (A)	The effect of increases in the minimum wage on employment.
4.3 (B)	The impact of block schedules and coaching on SAT scores.
4.3 (C)	Two articles on Parkinson's disease.

Problem 4.3 (A)

Raising the minimum wage

Whether or not to raise the minimum wage required to be paid to employees is a contentious issue.

1. In INFOTRAC, search on "minimum wage and case study." Select "Do minimum wages reduce employment? A case study of California, 1987-89," by David Card, *Industrial and Labor Relations Review*, October, 1992, v. 46, n. 1, pages 38-54.

   (a) Why are there few data points to analyze regarding minimum wage changes and the impact on employment and income?

   (b) Which groups does the author study? Why is it important to select comparable groups?

2. In INFOTRAC, search on "minimum wage and new jersey" and select from the available articles. See, for example, "Of magic, myth and the minimum wage" in *The Economist*, September 30, 1995, v. 336, n. 7934, page 94, or "The squabble over the minimum wage," by David R. Henderson, *Fortune*, July 8, 1996, v. 134, n. 1, pages 28-29.

   (a) Why would fast food establishments in New Jersey and Pennsylvania be comparable? How would they be different?

(b) Why are the authors criticized and what other evidence is available? For further information, see *National Bureau of Economic Research Working Paper* No. 5224, "The Effect of New Jersey's Minimum Wage Increase on Fast-Food Employment: A Re-Evaluation Using Payroll Records," by David Neumark and William Wascher: http://papers.nber.org/papers/W5224.

3. See the 1999 report "Assessing the Impact of Raising Wyoming's Minimum Wage" by the Wyoming Department of Employment: http://doe.state.wy.us/lmi/mw/mw.pdf.

   (a) What is a concern of those who oppose raising the minimum wage? See page 7.

   (b) What are two weaknesses of using CPS data? What are two for administrative data?

   (c) How is the sample restricted in the study? Why is this done?

   (d) What do the authors conclude? Do they prove that increasing the federal minimum wage reduces employment?

Problem 4.3 (B)

Studies at the College Board on Block Schedules and Coaching

Some students enroll in an SAT coaching program. School districts follow a variety of class schedules throughout the year. At http://www.collegeboard.com/research/home, search for keywords "block schedule" or "coaching" and select a study.

1. What groups are being compared? How were test takers "assigned" to groups?

2. What efforts are made in the study to make groups comparable?

Problem 4.3 (C)

Two articles on Parkinson's disease

Two articles on Parkinson's disease are available in INFOTRAC.

1. Search on "Parkinson's disease and twin sample" and select "Use of general practitioner computerised records to create a population based twin sample: pilot study based on Parkinson's disease," by C.H. Hawkes, A.M. Macdonald, and A.H.V. Schapira, *British Medical Journal*, December 6, 1997, v. 315, n. 7121, page 1510-1511. Search on "twins and ear infections," "twins and diabetes," or "twins and attention-deficit" for other twins studies. In what sense would two groups of twins, one with and one without a condition such as Parkinson's, resemble treatment and control groups in a randomized experiment?

2. Search on "Parkinson's disease and alternative" and select "Patients With Parkinson Disease and Alternative Therapy," by Mike Mitka, *JAMA, The Journal of the American Medical Association*, October 24, 2001, v. 286, issue 16, page 1961. If patients using alternative therapies have better health status than patients who do not use them, then can one conclude that the alternative therapies are effective? Why or why not? Explain what some of the confounding variables are.

Chapter 5

Probability

Introduction

Probability is concerned with the chance that an uncertain event happens. It describes the likelihood of possible results of a random, non-deterministic phenomenon. The terminology of sets can be used to denote individual outcomes, collections of outcomes, and, in the case of phenomena resulting in a numerical outcomes, intervals. Events are sets of outcomes. Introductory statistics textbooks describe set notation and operations on sets, including union, intersection, and complement.

The probability of an event can be defined in terms of the long-run relative frequency that it occurs when a random phenomenon happens. For example, the statement that a coin lands "heads up" half the time can be interpreted as a statement about the coin, how it is flipped, and what is expected over time. The process of measuring a random phenomenon also is important. If weights of several people are measured, then one might ask, what is the probability of recording a weight over 200 pounds? If one person is weighed several times, then questions about measurement error and change over time are relevant. Probability rules can be used to determine the probability of complex combinations of events.

The conditional probability of an event is the chance that an event happens given that something else has happened. Rules of probability apply to conditional probabilities, too. Bayes' theorem, which enables a reversal of conditioning, can be seen as one consequence of conditional probability. Tree diagrams and tables are useful for illustrating and computing with conditional probabilities. Many real applications involve comparing conditional probabilities and making predictions using reasoning formalized by Bayes' theorem.

Remarks

The mathematics of sets can be presented abstractly, but it also can be expressed in words and pictures, such as Venn diagrams and pie charts. It is important to know the rules of probability and to be able to use them in problems. If solutions to exercises are expressed in both symbols and words, then the likelihood of a correct solution increases. The abstract presentation of set theory and probability rules does have a purpose, though. In order to efficiently solve new problems, it is helpful to be able to relate the new story to the known rules and approaches, such as checking to see if the opposite event is easier to study. Further, probability problems can be converted into analogous problems involving tables, percentages, and fractions. Some people find it easier to reason correctly from a table of counts. Finally, practice is critical to being able to solve probability questions. An easy familiarity with probability rules and problem solving for most people comes through experience. See http://www.stats.gla.ac.uk/steps/glossary/probability.html and http://davidmlane.com/hyperstat/probability.html for on-line introductions to probability.

Outline

Section 5.1 illustrates basic rules of probability using Venn diagrams, pie charts, and tables. Problems in section 5.2 concern conditional probability and Bayes' theorem. Section 5.3 presents web sites related to interesting problems in probability. A couple of these problems are frequently used in introductory probability, but can be tricky to explain and illustrate. Numerical measurements of random events and the chances associated with numerical outcomes are discussed in chapter 6 on random variables and their probability distributions. Summaries of random variables are called statistics. They are random variables themselves and have probability distributions of their own, called sampling distributions in a repeated sampling framework. Sampling distributions are the focus of chapter 7. The Companion Website can be accessed at http://larsen.duxbury.com. Please enter the Serial Number from the back cover when prompted.

5.1   Probability Rules and Sets

Problem	Description
5.1 (A)	Pie charts used to illustrate long-term relative frequency.
5.1 (B)	Venn diagrams to illustrate rules for combining sets.
5.1 (C)	Questions about sets: Union, Intersection, Complement, and Difference.
5.1 (D)	On-line probability simulations involving dice.
5.1 (E)	Probability Rules and Card Games

Problem 5.1 (A)

Pie Charts for Representing Fractions, Percentages, and Probability

Go to http://www.shodor.org/interactivate/activities/piecharttool. Examine the three data sets, then return to "how I divide my 24 hour day."

1. Imagine printing a paper copy of the pie chart, attaching it to a dart board, and throwing a dart randomly at chart. Assuming you hit the chart, what is the probability that you hit ...

    (a) the area corresponding to sleep?

    (b) an area corresponding to something other than school or sleep?

    (c) an area corresponding to either chores, school, or homework?

2. See Wisconsin Energy Use Graphs at the University of Wisconsin, Stevens Point: http://www.uwsp.edu/cnr/wcee/keep/Mod1/Flow/graphs.htm.

    (a) How many total Btus (British thermal units) of Petroleum were used in Wisconsin in 1999?

    (b) Assume you have 1000 Trillion Btus of energy from Petroleum. If rates are the same as in 1999, how many and what percent would be used in each of six economic sectors?

46

Problem 5.1 (B)

On-line demonstration of Venn Diagrams for Sets and Probability

John Venn (1834-1923, see http://www.theory.csc.uvic.ca/~cos/venn/VennEJC.html) helped popularize diagrams known today as Venn Diagrams. Review http://www.cs.uni.edu/~campbell/stat/venn.html and http://infinity.sequoias.cc.ca.us/faculty/woodbury/Stats/Tutorial/Sets_Venn2.htm. See http://statman.stat.sc.edu/~west/applets/Venn.html or http://stat-www.berkeley.edu/users/stark/Java/Venn.htm, which depict two sets, A and B.

1. Arrange the blocks so that the intersection of A and B is 20 percent. Draw a picture or print the page to record your work. Rearrange the blocks so that the intersection of A and B is 10 percent.

2. Rearrange the blocks so that A and B are disjoint, representing mutually exclusive events A and B.

3. Point to one of the blocks, hold the mouse button down, close your eyes, and move the block around. Open you eyes. Verify the following equations: (a) $P(A \text{ or } B) = P(A) + P(B) - P(A \text{ and } B)$, (b) $P(A) = 1 - P(A^c)$, where A^c is A-complement or the opposite of A, (c) $P(\text{not}(A \text{ and } B)) = P(A^c \text{ or } B^c)$, (d) $P(\text{not}(A \text{ or } B)) = P(A^c \text{ and } B^c)$.

4. Events A and B are independent if $P(A \text{ and } B) = P(A)P(B)$. Move the block around so that this occurs. Draw or print the picture.

Problem 5.1 (C)

Dr. Math on sets: Union, Intersection, Complement, and Difference

Dr. Math has received and answered several questions concerning sets. Lets look at a few of these examples. Go to http://mathforum.org/library/drmath/sets/high_sets.html.

1. Read http://mathforum.org/library/drmath/view/61902.html on "Intersection, Difference, Union" of sets. See also http://mathforum.org/library/drmath/view/52389.html on "Intersection of Sets." Let $A = \{1, 2, 3, 4, \overline{5}, \overline{6}\}$, $B = \{-5, -4, -3, -2, -1, 0, 1, 2, 3, 4, 5\}$, and $C = \{1/6, 1/3, 1/2, 1, 2, 3, 6\}$.

   (a) Report the intersection of A and B, A and C, B and C, and A, B, and C.

   (b) What is the union of A and B, A and C, B and C, and A, B, and C?

   (c) List elements in A but not B, A but not C, B but not A, B but not C, C but not A, and C but not B?

2. See http://mathforum.org/library/drmath/view/52395.html on "Sets: Unions and Intersections." Let set C be the set of all capital letters with both curved and straight lines. How does C and the complement of C relate to sets A and B?

3. See "Unions and Intersections: Proving Sets" for a slightly more advanced question involving sets: http://mathforum.org/library/drmath/view/52446.html. Show, for any sets A and B,

   (a) the complement of (A intersection B) equals the complement of A union the complement of B.

   (b) the complement of (A union B) equals the complement of A intersection the complement of B.

4. See "Drawing Marbles" at http://mathforum.org/library/drmath/view/56491.html. Rework the problem assuming the jar contains three red, four blue, and six green marbles.

5. See "Independent and Dependent Events" at http://mathforum.org/library/drmath/view/56594.html. Write a sentence or two in your own words explaining why replacing the first marble into the box makes the events A and B independent, but not replacing it makes them dependent.

Problem 5.1 (D)

Probability Simulations involving Dice

Simulation is an effective tool for illustrating the meaning of long-run relative frequency and the frequency interpretation of probability. See http://www.shodor.org/interactivate/activities/racing/index.html. In this activity, two independent, fair dice are rolled and the sum is reported. If a sum corresponds to a lucky player's number, then the player advances toward a finish line. Set the length of the race equal to 1, so that there is a winner every time. Set the sum of 2 equal to player A, 3 equal to B, etc., so that each sum is associated with one player. Press "start the race." Change the number of times to 10000 and press Automatically run.

1. What fraction of the time did Player A win? What is the actual probability?

2. Repeat the calculations for the other players.

3. Press "change rules." Set all numbers equal to "none" except for 4 and 7. Now only 4 and 7 count and the contest is between players C and F. Run the race 10000 times. What fraction of the time did player C win? player F? How do the fractions relate to the actual probabilities? This problem relates to conditional probability, which is discussed in the next section.

4. The activity http://www.shodor.org/interactivate/activities/prob/index.html on spinners and dice also illustrates long-run relative frequency. Select the spinner and adjust it as you please. Run a simulation at least 100 times. How do the observed frequencies relate to the actual chances of each outcome?

Problem 5.1 (E)

Card games and probability rules

The exercise section of http://library.thinkquest.org/11506/learn.html discusses drawing cards from a standard 52-card deck. Probability rules are discussed at http://library.thinkquest.org/11506/prules.html.

1. See http://www.soyouwanna.com/site/syws/poker/poker.html for a description of a standard 52-card deck of cards and winning hands of Poker (see also http://www.gamingday.com/poker.shtml).

   (a) Suppose you draw two cards without replacement. That is, you two select a card, then do not replace it in the deck before drawing another. What is the probability that they have the same value? What is the probability that they have different values?

   (b) Suppose you draw two cards with replacement. That is, you replace your drawn card in the deck before drawing another. What is the probability that they have the same value? different values?

   (c) Suppose you draw 3, 4, or 5 cards. What are the probabilities?

   For more information: See high school probability (http://mathforum.org/library/drmath) at Dr. Math. Select for search option for "just High School Probability" and enter a key word such as cards or poker. Study the questions and answers. Some answers use combinatorics, which are discussed in section 6.2.

2. See http://www.gamingday.com/blackjack.shtml about the game of blackjack, or 21. An ace counts as 1 or 11 points. Tens and face cards (jacks, queens, kings) count as 10 points. The goal is to possess cards with values adding to as close to 21 as possible, but not more. Cards are dealt without replacement.

   (a) What is the probability that the total is 17 on two cards? 18? 19? 20? 21?

   (b) Assume that the sum of two cards is 17. What is the probability that the next card dealt increases the sum to 18? to 19? to 20? to 21? over 21?

   (c) Assume that the sum of two cards is 18. What is the probability that the next card dealt increases the sum to 19? to 20? to 21? over 21?

5.2   Conditional Probability and Bayes' Theorem

Problem	Description
5.2 (A)	Interactive illustration of conditional probability using Venn diagrams.
5.2 (B)	Tree diagrams, conditional probability, and tests for Lyme disease.
5.2 (C)	Bayes' theorem calculations and sensitivity and specificity of tests for influenza.
5.2 (D)	A couple of articles in which Bayes' theorem is applied.

Problem 5.2 (A)

Conditional Probability and Venn Diagrams

See one of the following applets: http://statman.stat.sc.edu/~west/applets/Venn.html or http://stat-www.berkeley.edu/users/stark/Java/Venn.htm. The pictures depict two sets A and B.

1. Move the blocks around. Place them so that they overlap. You can draw or print this page.

   (a) What is $P(A|B) = P(A \text{ and } B)/P(B)$?

   (b) What is $P(B|A) = P(A \text{ and } B)/P(A)$?

   (c) What is $P(A^c|B) = P(A^c \text{ and } B)/P(B)$?

   (d) What is $P(A|B^c) = P(A \text{ and } B^c)/P(B^c)$?

2. Move the blocks around so that $P(A|B)$ increases. Answer the questions above.

3. When A and B are kept the same size and $P(A|B)$ increases, does $P(B|A)$ also have to increase? Why or why not?

Problem 5.2 (B)

Conditional Probability and Tree Diagrams

Conditional probabilities can be displayed using tree diagrams. See the web site http://statman.stat.sc.edu/~west/applets/tree.html for a demonstration tool.

1. Enter numbers $\begin{array}{|c|c|} \hline 80 & 12 \\ \hline 10 & 15 \\ \hline \end{array}$ into the table and press compute.

   (a) What is the probability that a randomly selected student passes the first exam? the second exam?

(b) Given that a student passed the first exam, what is the chance that the student passed the second exam? failed the second exam?

(c) Given that a student passed the first exam, what is the chance that the student passed the second exam? failed the second exam?

(d) Are passing the first and second exams independent? Why or why not?

2. Enter numbers $\begin{array}{|c|c|} \hline 98 & 42 \\ \hline 14 & 6 \\ \hline \end{array}$ into the table and press compute. Answer the questions asked above.

3. Lyme disease tests are described at http://www.caes.state.ct.us/PlantScienceDay/2002PSD/Ticks.htm. Read the article, especially the second to last paragraph. Assume ten percent of the deer in the state of Connecticut are infected with Lyme disease. The sensitivity of a test is the probability that a test is positive given that the animal is infected. The specificity of a test is the probability that a test is negative given that the animal is not infected. Pick one of the tests and record its sensitivity and specificity.

(a) What percent of deer test positive? You can make a table or a tree diagram. For a table, assume you have 1000 deer.

(b) What is the predictive value positive? That is, what fraction of deer that test positive actually are infected?

(c) What is the predictive value negative? That is, what fraction of deer that test negative actually are not infected?

(d) Repeat the above exercise for a different test or assuming a twenty percent infection rate.

Further Activity: Express characteristics of the labor force in two-by-two tables and tree diagrams. You can have more than two branches at each level. See http://www.bls.gov/cps/home.htm#tables for the national counts and problem 3.1 (B) for some discussion. See also examples of trees at http://mathforum.org/library/drmath/view/56601.html. Other information on Lyme disease and diseases in deer were studied in problem 3.1 (C).

Problem 5.2 (C)

Bayes' Theorem – A Demonstration of Calculations

Thomas Bayes (http://www-groups.dcs.st-and.ac.uk/~history/Mathematicians/Bayes.html) was an English clergyman and mathematician in the Eighteenth century. One of his papers, published posthumously, presents what today is referred to as Bayes' Theorem.

1. See http://statman.stat.sc.edu/~west/applets/bayesdemo.html. Move the square in the center so that events A1 and A2 have probabilities less than 0.15 and 0.20, respectively. The square B occupies 25% of the total area. Assume that the probability is uniformly distributed over the total area. In the following, you may make tree diagrams or a table with 10,000 total counts.

(a) What is the probability of being in square A1 given that you are in square B? Answer for A2, A3, and A4, too.

(b) What is the probability of being in the intersection of A1 and B? A2 and B? A3 and B? A4 and B? Does this add to 0.25?

(c) What is the probability of being in B given that you are in A1? A2? A3? A4? Verify calculations using Bayes' Theorem.

2. See http://statman.stat.sc.edu/~west/applets/bayescalc.html. Choose a problem, such as one of those mentioned in problem 5.2 (B) or search on INFOTRAC using keywords "sensitivity" and "specificity." One article is "An office-based approach to influenza: clinical diagnosis and laboratory testing" by Norman J. Montalto, *American Family Physician*, January 1, 2003, v. 67, issue 1, pages 111-118. Define event B, such as having a disease or being unemployed. Define events A1, A2, etc., that divide the population or sample exhaustively into mutually exclusive groups. Change the number of events to match the number of categories represented by the A events. Enter the appropriate numbers. You may make a tree diagram if it is helpful.

(a) What is the probability of event B and event A_k, where k indexes the categories? Describe these events in words.

(b) Verify the probability of event B. Describe this event in words.

(c) Verify using Bayes' theorem the probability of event A_k given B. Describe these conditional probabilities in words.

Problem 5.2 (D)

Some applications in articles of Bayes' Theorem

Some applications of Bayes' Theorem are presented in articles referenced below.

1. In INFOTRAC, search on "Bayes, Thomas" and select the article "The mathematics of making up your mind," by Will Hively. *Discover*, May, 1996, v. 17, n. 5, pages 90-97.

(a) Read the analogy involving the bowl with black and white balls on page 93. Imagine a bowl with three balls in it. Two are of one color and one is other the other. Let A be the event that the majority is black. If you assume an *a priori* even distribution of colors, what is your $P(A)$?

(b) Let B be the event that you draw a black ball. Suppose that you draw a black ball. What is $P(A|B)$? What is $P(A^c|B)$, where ^{c} denotes complement?

(c) Suppose you draw a second black ball. What is the probability of A given this second event?

(d) In what way does subjective opinion enter into a Bayesian statistical analysis?

2. In INFOTRAC, search on "Bayes' Theorem" and retrieve the article "Chest pain with normal coronary arteries," by Ezra A. Amsterdam *et al.*, in *Patient Care*, March 15, 1997, v. 31, n. 5 pages 43-49. At the end of the article is a section discussing the use of Bayes' Theorem. Let A be the event that the patient has coronary artery disease (CAD). Let B be the event that the patient tests positive on a stress test.

(a) For one risk group, as described in the appendix, what is $P(A)$? Define your risk group.

(b) What is $P(B|A)$ and $P(B|A^c)$? Describe these events in words.

(c) Determine $P(A|B)$ and $P(A|B^c)$. What do these events mean? You may use a tree diagram, the Bayes' calculator of the previous problem, or a table of counts.

5.3   A Couple of Probability Examples

Problem	Description
5.3 (A)	The Birthday Problem.
5.3 (B)	The Monte Hall Problem, or Let's Make a Deal.

Problem 5.3 (A)

The Birthday Problem

Read the description of the problem at http://www.mste.uiuc.edu/reese/birthday. In the problems below, we will assume that all birthdays are equally likely and independent of one another.

1. What is the probability that two people have the same birthday?

2. What is the probability that among three people at least two have the same birthday? How about for four people?

3. How many people need to be in a room before the probability that at least two have the same birthday is over fifty percent?

   (a) Lets see a simulation. Select the graphical applet at http://www.mste.uiuc.edu/reese/birthday. Run the applet 100 times each for 15, 20, 25, and 30 birthdays.

   (b) Click on the explanation link at the bottom. Write an explanation in your own words. Define any symbols you use. See also http://mathforum.org/dr.math/faq/faq.birthdayprob.html and http://www.math.hmc.edu/funfacts/ffiles/10001.6.shtml.

Problem 5.3 (B)

The Monte Hall Problem, or "Let's Make a Deal"

Read the description of the problem at one or more of http://www.mste.uiuc.edu/reese/monty/monty.htm, http://statman.stat.sc.edu/~west/applets/LetsMakeaDeal.html, or http://www.shodor.org/interactivate/activities/monty3/index.html (click on what?).

1. What is your opinion? Should you switch after being shown a door without the prize or stay with your original pick? Run one of the simulations a few times.

2. Make a probability tree diagram. Assume you choose the first door and play the strategy of staying with this pick. The prize could be under any of the doors with probability 1/3 each. For each prize location, which doors can be opened without revealing a prize? If there is more than one option, assume that they are opened with equal probability. What is the probability of winning given that you stay? Repeat the tree diagram (if you need to convince yourself) for other initial choices and strategies.

3. Run the simulation 100000 times. See http://www.shodor.org/interactivate/activities/monty/index.html. What percent of the time do you win if you stay? if you switch?

4. Make it so that there are ten doors. Monte Hall opens all but one unchosen door before the decision to switch or stay is implemented. Run the simulation 100000 times. What percent of the time do you win if you stay? if you switch?

5. See discussions at http://mathforum.org/library/drmath/view/52143.html and http://www.dartmouth.edu/~chance/course/topics/Monty_Hall.html. Write in your own words why one strategy is better than another.

Chapter 6

Random Variables and Probability Distributions

Introduction

Random variables record numerical outcomes of random phenomena. The probability distribution of a random variable indicates the chances, or probabilities, of the possible outcomes. If the random variable is discrete, then a list of possible values and their probabilities defines the probability distribution. Sometimes the probabilities corresponding to a list of outcomes can be indicated efficiently by a formula. If the random variable can take on values in a continuum, the relative likelihood of values is represented by a probability density function. The probability of observing a value in an interval is given by the area under the density function above the interval.

The probability distribution of a random variable can be thought of as a model of a random phenomenon. The model might describe exactly the chances of the possible outcomes, or it might be an approximation that is simpler to describe than the real process. Commonly used random variables include the binomial, geometric, Poisson, normal, and uniform random variables. Probability distributions and densities are often represented by bar graphs and density plots, respectively. The density plots are like smoothed versions of histograms. The overall shape, as well as center, spread, and skewness, are of interest. The mean, median, interquartile range, and standard deviation of a random variable can be reported. A discussion and definitions can be found at http://www.stats.gla.ac.uk/steps/glossary/probability_distributions.html. A gallery of distributions with technical details is presented at http://www.itl.nist.gov/div898/handbook/eda/section3/eda366.htm.

Remarks

The idea of a random variable is very general. Any aspect of a random phenomenon that can be recorded numerically is a random variable. The probability distribution depends on what exactly is being measured and how it is being recorded numerically. Specific models, such as binomial and normal, are useful in many situations, but certainly do not apply to most. In fact, it is often the case that random variables do not have exactly known probability distributions. Chapters 1, 2, and 3 presented methods for summarizing observed values, whereas this chapter concerns models for measurements, which might be made in the future. The methods of descriptive statistics can be thought of applying to sample values. In contrast, probability models

53

describe a theoretical possibility or a population distribution.

Outline

Section 6.1 presents examples of general random variables. Sections 6.2 and 6.3 provide illustrations of discrete random variables. Section 6.2 focuses on the binomial random variable and its probability distribution. Section 6.3 contains problems on the geometric and Poisson random variables. Section 6.4 concerns the normal distribution Section 6.5 uses the normal distribution to approximate the binomial probability distribution. Section 6.6 contrasts independence and dependence for random variables.

6.1   General Random Variables

Problem	Description
6.1 (A)	Center for Disease Control growth charts.
6.1 (B)	Random variables from dice and forest fire simulations.
6.1 (C)	Roulette and lottery outcomes.

Problem 6.1 (A)

General random variable: CDC clinical growth charts

The Center for Disease Control (CDC) clinical growth charts can be used to monitor the growth of infants and children. See http://www.cdc.gov/growthcharts. One set of charts presents weight, length, and head circumference by age and weight by length for boys and girls from birth to 36 months of age.

1. For a randomly selected 18-month-old boy, what are the (approximate) median length, weight, and head circumference? What are they for a girl?

2. For boys and for girls, what are the interquartile ranges for these random variables?

3. What are the 5th and 95th weight percentiles for a 80cm boy? a 90cm boy? How about for girls?

Problem 6.1 (B)

General random variable: Racing with Coins and Dice-controlled Fires

1. A dice game was studied in problem 5.1 (D). See http://www.shodor.org/interactivate/activities/racing. In this activity, two independent, fair dice are rolled. The applet reports the sum.

   (a) You roll the dice until a 4 or 7 is recorded. What percent of the observations are 4? What is the probability of a 7? Simulation: Set 4 to Player A and 7 to Player B. Run 10000 times. What is the actual probability distribution?

   (b) Suppose you record the maximum of the two values. What is the probability distribution?

   (c) Suppose you record the absolute difference of the two values. What is the probability distribution?

2. The activity http://www.shodor.org/interactivate/activities/fire1 simulates a fire using random numbers. Read the What? How? and Why? pages. Set the probability that a neighboring tree catches on fire to 1/6, or to some number other than 0 and 1. After a simulated fire burns, you can re-grow the forest.

(a) Start the fire at the tree in the fourth colum and fourth row. What is meant by the probability distribution of the number of trees that burn? Simulate it a few times.

(b) What other random variables can be recorded based on single fires such as those in part (a)?

(c) How would you simulate the probability distribution of number of fires before nine trees burn? *For your information*: The directories firealt and fire2 for other fire simulations.

Problem 6.1 (C)

General random variables: Roulette and lottery outcomes

1. The game of roulette is played by placing a chip on one or a combination of 38 numbers on the board. Eighteen numbers are red, eighteen are black, and two are green. A ball is spun around a spinning wheel and where it stops determines the winning number. See http://www.gamingday.com/roulette.shtml and http://www.casinocenter.com/htp/roulette for more information.

   (a) Suppose you bet on a single number and record how much you win. A payoff of 35:1 means that you receive $35 if you win, but pay $1 if you lose. What is the probability distribution of this random variable?

   (b) What are the expectation and standard deviation of the payoff amount in the previous part?

   (c) Bets on two numbers have a payoff of 17:1. Bets on red or black have a payoff of 1:1. What are the probability distributions of the payoffs for these bets?

   (d) What are the expectation and standard deviation of the payoff amounts in the previous part?

2. The North American Association of State & Provincial Lotteries (http://www.naspl.org) provides links to lotteries in North America (http://www.naspl.org/uscanada.html), including the Illinois lottery (http://www.illinoislottery.com). Odds and prizes for various games are listed at http://www.illinoislottery.com/win.htm. An odds of winning of $a : b$ means that the chance of winning is $a/(a+b)$. For the Illinois lottery game Lotto, assume a $2 million grand prize.

   (a) For the Illinois lottery game Lotto, what is the probability distribution?

   (b) What is the expected return for one play?

   (c) What would be the expected return for one play if the grand prize were $10 million?

6.2 Binomial Random Variables

Problem	Description
6.2 (A)	Success in sports and the binomial distribution.
6.2 (B)	Success in sports and binomial probability calculations.
6.2 (C)	Assumptions of the binomial distribution and sports.

Problem 6.2 (A)

A binomial Probability Applet and Sports success

Applet http://www-stat.stanford.edu/~naras/jsm/example5.html depicts the binomial probability distribution. Look up success percentages related to a sport. Record percentages for two or three players of different abilities in a few categories.

55

- Basketball: http://www.nba.com. Select the link to Statistics. Find the NBA league leaders in Field Goal, Free Throw, and 3 Point Field Goal Percentages.

- Golf: http://sports.espn.go.com/golf/statistics. Find the PGA leaders is percent of greens in regulation, sand saves, and driving accuracy. The number of holes per eagle is not a percentage.

- Baseball: http://www.mlb.com: Select Stats. Find the MLB leaders in percentages for batting average (this is a percent) and slugging percentage.

1. Suppose that a player makes 18 attempts (18 shots of one type in basketball, 18 holes of golf, 18 at-bats in baseball) and that the binomial model is appropriate, i.e., independent attempts, constant chance of success on each attempt. For one cateogory, as close as possible, make binomial probability histograms for a couple of players. Write a couple of sentences comparing the histograms.

2. Make probability histograms for results in two categories. Write a couple of sentences comparing the distributions.

3. Suppose players make 36 attempts. Make probability histrograms. How do the distributions in the previous parts change? Write a couple of sentences about one of the situations.

Problem 6.2 (B)

A binomial Probability calculations and Sports success

See http://www.stat.sc.edu/~west/applets/binomialdemo.html. This applet allows you to view the probability of individual outcomes and cumulative probabilities. Choose a sport and an activity. Choose two players. Suppose each player is going to do the activity 18 times, the binomial distribution is appropriate, and the proportion p is equal to the proportion reported on the web site.

1. For the player from among those you selected previously with a percent success closest to 50 percent, compute the mean and standard deviation (SD) of the binomial distribution. Let u be the integer closest to the mean plus one SD. Let l be the integer closest to the mean minus one SD.

2. What is the probability that each player has u successes out of 18 trials? Which player has the higher probability? Explain why this result makes sense.

3. What are the probabilities that the players have l successes out of 18 trials?

4. What is the probability that each player has between l and u successes inclusive? That is, what is $P(l \leq X \leq u)$?

5. What is the probability that each player has between l and u successes exclusive? That is, what is $P(l < X < u)$?

6. Suppose that one player does one activity 180 times. Assume that the player rests enough so that the chance of success does not decrease over time. Compute the mean and SD. Find integers u and l closest to the mean plus and minus one SD. Using the binomial model, what is the probability of exactly u and l successess? Why are they so much smaller than the probabilities computed before?

Problem 6.2 (C)

Binomial distribution assumptions and golf

In which of the following studies is a binomial distribution useful as a model? If the binomial distribution is not useful, then what aspect of the model is inappropriate? Go to http://sports.espn.go.com/golf/statistics.

1. The number of putts per hole is the number of hits of the ball with the putting club (a particular golf club) it takes to get the golf ball into the hole after the ball is one the green, the area near the hole. This variable records the number of strokes or hits necessary to get the ball in the hole. There are 18 holes per round, so 10 rounds equals 180 holes. If you assume a golfer plays 18 holes of golf, can this variable have a binomial distribution? Why or why not?

2. Getting to a green in regulation means that the players hits the ball in few enough shots onto the green, the area near the hole. Suppose there are 18 holes. Can the number of times the player gets to the green in regulation be considered a binomial random variable? Answer why or why not for two situations.

   (a) Assume that the number of shots allowed in regular is adjusted for the difficulty of the holes so that the chance of getting to the green in regulation is the same for every hole.

   (b) Assume that the adjustment for the difficulty of the hole is not perfect, so that the chance of getting to the green differs by hole.

3. Consider variables associated with another sport. Describe two that can be modeled using the binomial and two that cannot. Be specific about how, when, and how often the variables are measured.

6.3 Geometric and Poisson Random Variables

This section presents problems related to geometric and Poisson random variables.

Problem	Description
6.3 (A)	Poisson random variables and hockey.
6.3 (B)	Geometric random variables, color deficient vision, and blindness.

Problem 6.3 (A)

Poisson random variables and goals in hockey

Models for goals in the game of hockey sometimes include the Poisson distribution as one component. The Poisson distribution is named after Simon-Denis Poisson, who published results concerning the distribution in 1837; see http://www-gap.dcs.st-and.ac.uk/~history/Mathematicians/Poisson.html. In INFOTRAC, search on "Poisson" and select the article "Late-Game Reversals in Professional Basketball, Football, and Hockey" by Paramjit S. Gill in *The American Statistician*, May, 2000, v. 54, issue 2, pages 94-99. Another article that uses the Poisson distribution in an analysis of Hockey scores is by G.M. Mullet (1977), "Simeon Poisson and the National Hockey League," *The American Statistician*, volume 31, issue 1, pages 8-12. Hockey games consist of three 20-minute periods. Gill (2000) reports that the number of goals scored in the first two periods by the home and away teams can be modelled using separte Poisson distributions. The mean, which is equivalent to the rate λ, is 1.770 for the home team and 1.694 for the away team. Mullet (1977) considered models for goals for a team and goals by an opposing team when the team played at home and away. For example, the Boston Bruins in the 1973-74 season scored an average 4.95 goals per game (three periods, gpg) at home and their opponents scored an average of 2.43 gpg. The Philadelpia Flyers playing on their home ice averaged 3.90 gpg, whereas their opponents averaged 1.92 gpg. Boston and Philadelphia were the teams with the best records in the 1973-74 professional hockey season.

1. Simulation: http://www.math.csusb.edu/faculty/stanton/m262/poisson_distribution/Poisson_old.html. Choose a value of the parameter λ, the mean of the Poisson distribution. The value could correspond to average numbers of hockey goals. Simulate 20 repetitions of observing the value of a Poisson random variable. Keep the nubmer of repetitions at 1 and hit the "fish" button 20 times. Change the number of repetitions to 20 and perform 180 more repetitions, so that you have 200 in total. Estimate the probability that $X = 0$, $X < 2$, and $X > 5$ based on the histogram of simulated values. Repeat this for a different value of the parameter. Write a sentence comparing the results.

2. Calculation: http://www.stat.sc.edu/~west/applets/poissoncal.html. For two different values of the Poisson parameter, calculate the probability that $X = 0$, $X < 2$, and $X > 5$. Check the results using the Poisson probability formula. How well did the 200 simulated observations from part (a) estimate these probabilities?

3. Expected outcomes: http://www.anesi.com/poisson.htm. The expected number of trails that yield a certain number of events is the number of trials times the probability of having that number of events. (This actually is related to the mean of a binomial variable.) Suppose the average number of events on a single trial is 4 and the Poisson model is appropriate. How many times out of 1000 trials would you expect to have no occurrences of the event? less than 2? more than 5? How do these expected numbers change when the average number of events on a single trial is 3? Write a sentence contrasting the results based on the two averages.

Problem 6.3 (B)

Geometric Random variable and looking for rare traits

A random variable with a geometric probability distribution records the number of trials before a success. The probability mass function (PMF), representing the probability distribution, of a geometric random variable http://links.math.rpi.edu/devmodules/probstat/concepts/html/geometric.html. is decreasing.

1. Compare the PMF when the rate of success is 0.2 versus 0.5. Which distribution has the larger mean? That is, which distribution on average takes longer to produce a success?

2. People who have color deficient vision do not see the full spectrum of colors. According to http://www.toledo-bend.com/colorblind, about ten percent of men from European origins have color deficient vision. See http://www.nlm.nih.gov/medlineplus/ency/imagepages/9962.htm for color vision tests. Suppose you want to interview men of Europen decent who have color-deficient vision, but randomly sample men from this population at large.

   (a) What is the chance that you find the first person with color deficient vision on your second try? on your third interview?

   (b) What is the probability that you will need to sample more than five people?

   (c) How many people would you expect to need to sample to find one individual with this condition?

3. According to the American Foundation for the Blind (http://www.afb.org, see "blindness and low vision," then "statistics," then "statistics for professionals"), 0.5 percent of Americans are legally blind. The National Center for Health Statistics (http://www.cdc.gov/nchs) is cited as source of this estimate. If this is correct and you randomly sample people in the U.S., answer the questions of the previous part.

6.4 Normal Random Variables

The normal distribution is discussed at http://davidmlane.com/hyperstat/normal_distribution.html and http://www.stat.wvu.edu/SRS/Modules/Normal/normal.html. A historical note on the normal density can be found at http://www-gap.dcs.st-and.ac.uk/~history/Curves/Frequency.html. Problems in section 6.4 on the normal distribution and density are listed in the table below.

Problem	Description
6.4 (A)	Chest sizes of Scottish militiamen in the early 19th century.
6.4 (B)	Understanding the Normal Density and the empirical rule.
6.4 (C)	Using the normal density to model the effect of coaching on SAT scores.
6.4 (D)	The impact of a small gender bias, studied using normal distributions.

Problem 6.4 (A)

Chest sizes of Scottish militiamen in the early 19th century

DASL displays a histogram of chest sizes (the distance around the chest) of 5738 men in the Scottish army in the early 1800s. See http://lib.stat.cmu.edu/DASL/Stories/ChestsizesofMilitiamen.html for the description and http://lib.stat.cmu.edu/DASL/Datafiles/MilitiamenChests.html for the data.

1. Does the histogram match a normal density exactly? If not, how do the data deviate from being a sample from a normal distribution. Hint: consider the issue of "rounding" in addition to the picture.

2. Using the summary data, compute the mean (between 33 and 48) and standard deviation (around 15/4). Note that $n = 5738$ and many men have the same measurement.

3. Using your estimates and modeling the data using a normal distribution, what percent of the men would be expected to have chest bigger than 43.5 inches? What percent of the men actually do?

4. What is the 30th and 70th percentile of a normal distribution with the mean and standard deviation that you calculated? What are the actual 25th and 75th percentiles for these data?

Problem 6.4 (B)

Understanding the normal density and the empirical rule

Applets http://www-stat.stanford.edu/~naras/jsm/NormalDensity/NormalDensity.html and http://statman.stat.sc.edu/~west/applets/normaldemo1.html illustrate the normal density.

1. How does the normal probabililty density curve change as the as the mean changes? Change the mean to various values and write a summary. Note that the vertical axis on the curve remains the same. See

2. How does the normal probabililty density curve change as the standard deviation changes? Change the standard deviation to various values and write a summary.

3. The probability that a normal random variable takes a value in an interval is determined by area under the normal curve above that interval. See http://www.stat.sc.edu/~west/applets/normaldemo2.html. Using the website, for a standard normal random variable (mean 0, standard deviation 1),

   (a) what is the probability that the variable has a value less than 0.3? -0.3? 0?

(b) what is the probability that the variable has a value greater than 0.2? -0.2? 0?

(c) what is the probability that the variable has a value between ±0.3? ±0.2? -0.5 and 1.2?

(d) what is the 25th percentile? the 75th percentile? the interquartile range?

4. Derivation of the empirical rule:

(a) What percent of the observations from a normal distribution are within one, two, and three standard deviations of the mean? See http://www.stat.sc.edu/~west/applets/empiricalrule.html.

(b) Do the percentages change when you change the mean and SD? Experiment and write a summary.

(c) Summarize the rationale for the 1.5 IQR rule: http://exploringdata.cqu.edu.au/normdist.htm.

Problem 6.4 (C)

Coaching and the SAT: using the normal density for calculations

Normal distributions are sometimes appropriate for modeling SAT math and verbal scores. Look up means and standard deviations (SDs) in http://www.collegeboard.com/research/html/rr9806.pdf, College Board Report No. 98-6 (June, 1998), "Effects of Coaching on SAT I: Reasoning Scores" by Donald Powers and Donald Rock. See table 2 on page 11. For this exercise, we will pretend that these values are the means and SDs of the populations. There are summaries of pre-test and post-test scores, as well as gain scores, which are positive for students who increase their scores on a second attempt. You can record gain score values, too.

1. What is the probability of getting a score below 400? Report this probability for at least two tests or groups. Draw normal densities by hand on ONE graph to illustrate the probabilities.

2. What is the probability of a score between 450 and 575? Report for at least two tests or groups. Illustrate the probabilities on ONE graph (or on two graphs, one above the other).

3. What are the eightieth and the fortieth percentiles for at least two tests or groups? Draw ONE graph (or two graphs, one above the other) to illustrate the percentiles.

4. The gain scores are the differences between the post- and the pre-test scores. For the general population we'll assume they follow a normal distribution. What is the probability that a gain score is positive? What are the eightieth and the fortieth percentiles for gain scores?

Problem 6.4 (D)

The impact of a small gender bias on high-level applicants

In this problem the evaluation scores of females and males applications are assumed to follow normal distributions. The distributions have the same standard deviation (SD), but differ in mean. The scores might be based on exams, grades, interviews, or a combination of several measurements, which are quantified into a single numerical value. See http://www.ruf.rice.edu/~lane/stat_sim/group_diff.html. One normal curve represents the female applicants; the other the males. It is assumed that the difference in means is due to a bias in the evaluation process. What impact does a small bias have on the comparison of highly qualified applicants?

1. Set the mean for the red curve at 55 (the mean for blue curve is 50). Set the cutoff value at 70. Applicants with scores above 70 get hired. Explain what is displayed in the pictures. If the SD for each curve is 10 points, what percent under the blue and red curves have scores above 70 points? What is the ratio of these probabilities (red/blue)?

2. Change the mean of red curve to 53 and then to 57. How do the probabilities and the ratios change?

What else is here? The percent of the variability explained by the group difference is the ratio of two terms. The numerator is the square of half the distance between the means: $((\mu_1 - \mu_2)/2)^2$. The denominator is the numerator plus the variance within a group: $10^2 = 100$. If the difference is 10, the percent of the variabiltiy is $5^2/(5^2 + 10^2) = 25/125 = 0.20$, or 20%. The decomposition of variability into differences between groups and the variability within groups is the basis of analysis of variance.

6.5   Normal Approximation to the Binomial Distribution

Problem	Description
6.5 (A)	A study of the normal approximation to the binomial.
6.5 (B)	Blood donation and the normal approximation to the binomial.

Problem 6.5 (A)

Normal Approximation to the binomial Distribution

The normal distribution can be used to approximate the binomial distribution when the sample size n of the binomial is large. The approximation is effective because for large n the normal density curve can be matched closely to the binomial probability histogram. See http://stat-www.berkeley.edu/~stark/Java/BinHist.htm or http://www.ruf.rice.edu/~lane/stat_sim/normal_approx. Use these tools to judge the accuracy of the normal approximation. Both of these applets use a continuity correction for the normal approximation.

1. For a sample of size 10, 20, or 40, which numbers of included in the following ranges?

   (a) Less than 30 percent (excluding 30 percent) of the trials.

   (b) Less than or equal to 30 percent of the trials.

   (c) 20 percent to 60 percent (inclusive) of the trials.

   (d) 70 percent to less than 90 percent of the trials.

2. For the above situations (three sample sizes, four scenarios), what interval is used in the normal approximation with a continuity correction?

3. For $p = 0.50$, what are the exact binomial probabilities and what are the normal approximation values for the above situations? In which cases is normal approximation accurate?

4. For $p = 0.20$, what are the exact binomial probabilities and what are the normal approximation values for the above situations? In which cases is normal approximation accurate?

Problem 6.5 (B)

Normal approximation to the binomial distribution: blood donation and testing

The American Association of Blood Banks (http://www.aabb.org) answers frequent questions about blood on its web site. See the link called "All About Blood."

1. Suppose 5% of the adult population of a large city dontes blood in a calendar year. If you randomly select 300 unrelated adults, what is the probability that over 20 donated blood? under 10? 12 to 18?

2. Suppose you randomly select 200 unrelated people from a large community. What is the chance that less than thirty percent have type A+ blood? O+? B+?

3. According to the Center for Disease control (http://www.cdc.gov/ncidod/diseases/hepatitis), 1.8 percent of Americans are infected with Hepatitis C. If you randomly select individuals to test, how many should be selected before for the normal approximation to be reasonable?

Note: The populations that donate blood and that are at high risk for Hepatitis C are very different from each other and from the general U.S. population.

6.6   Independent Versus Dependent Random Variables

Problem	Description
6.6 (A)	Are Math and Verbal SAT scores independent?
6.6 (B)	Joint Distribution of SAT Math and Verbal Scores

Problem 6.6 (A)

Are math and verbal SAT scores independent?

The College Board reports mean and standard deviations for Math, Verbal, and total SAT scores. If Math and Verbal SAT scores are independent, then the variance of the sum of individual scores should be the sum of the variances. The summary statistics are based on the entire population (tens of thousands of students) of SAT test takers for a given year. See http://www.collegeboard.com/counselors/hs/sat/scorereport/scoredata.html.

1. What is the relationship between the mean total SAT score and the Verbal and Math means?

2. Is the variance of the sum of individual scores equal to the sum of the variances? Is the standard deviation of the sum more or less than the square root of the sum of the variances? If its more, then scores are positively correlated. If its less, they are negatively correlated. Comment on your results.

3. The variance of a sum of two dependent random variables, V and M, is $\text{var}(V + M) = \text{var}(V) + \text{var}(M) + 2\text{cov}(V, M)$, where $\text{cov}(V, M) = \text{cor}(V, M)\text{SD}(V)\text{SD}(M)$ is the covariance which is equivalent to the correlation times the SDs of the two variables. What is the correlation between SAT Verbal and Math scores?

Problem 6.6 (B)

Jointly distributed SAT Math and Verbal scores: the bivariate normal distribution

This problem examines the normality assumption use in the example of the previous problem.

1. The empirical rule states that within approximately 1-2-3 SDs of the mean is 68-95-99.7% of the area under the normal density.

   (a) Convert the empirical rule statements into six statements about normal distribution percentiles. For example, the mean minus one SD is the sixteenth percentile, because (100-68)/2 = 32/2 = 16% of the area under a normal curve is less than this value.

   (b) Compare the six values for the Verbal score percentile ranks to their expected values according to the normal model. Make a table and write a sentence summarizing the comparison. In some cases when reading scores from the table, it will be necessary to interpolate or approximate.

(c) Repeat the previous question for Math scores.

2. The interquartile range is the distance from the twenty-fifth to the seventy-fifth percentile.

   (a) Find these percentiles for the standard (mean zero, SD one) normal.

   (b) Compare the two values for the Verbal score percentile ranks to their expected values according to the normal model. Write a sentence summarizing the comparison.

   (c) Repeat the previous question for Math scores.

3. Draw a picture describing the joint distribution of SAT scores assuming they follow a bivariate normal distribution. The steps below guide you through the creation of the picture.

   (a) The range of scores for both exams in 200-800; draw axes for a scatterplot with these ranges. Indicate where the means of the two variables are.

   (b) Draw a box on the scatter plot that depicts where most of the data values would be. About 95 percent of the values for both variables are going to be within two SDs of their means.

   (c) Draw the regression line for predicting Math based on Verbal scores and the line for predicting Verbal based on Math scores. When drawing one of these lines, it might be helpful to rotate the page 90 degrees counter-clockwise. The regression equations were calculated in problem 2.2 (G). Remember: the regression lines go through the point defined by the two means.

   (d) Draw an ellipse inside the box that touches the sides of the ellipse where the regression lines touch the box. If the scores follow a bivariate normal distribution, then most of the scores will be inside this ellipse.

   Further Study: See part (d) of the sixth free-response problem of the 2000 AP Statistics Exam: http://apcentral.collegeboard.com/members/article/1,3046,152-171-0-8357,00.html.

Chapter 7

Sampling Distributions

Introduction

Statistics are functions, or summaries, of random variables. Once the random variables are measured, i.e, the data are collected, the value of the statistic can be computed and reported. The distinction between a statistic as a general formula and its value for a particular data set is formal, but useful when introducing the idea of statistical inference. An example of a statistic is the sample average, $\bar{X} = \sum_{i=1}^{n} X_i/n$, where X_i is the measurement of random variable X on individual i. No matter who is selected for the sample, $\bar{X}$ represents summarizing the data by computing the sample average. Once $n = 10$ heights of sixth grade girls are measured, their average height is $\bar{x} = \sum_{i=1}^{n} x_i/n$, where x_i is the height of girl i. If another group of ten girls were measured, $\bar{x}$ most likely would be different.

Statistics are random variables themselves and have probability distributions of their own. If one imagines repeating the data gathering or measuring process over and over again independently and under identical conditions, the probability distribution of a statistic is referred to as its sampling distribution. A statistic can be any numerical summary of the data, including measures of center, spread, skewness, and extremeness. All statistics, from the repeated sampling perspective, have sampling distributions. In a few special cases, sampling distributions are known and well-understood distributions. For example, if n independent trials are made with probability of success p on each trial, the number of success has a binomial probability distribution. The binomial distribution is sampling distribution for the number of successes. In most cases, however, sampling distributions are not known exactly and must be studied through simulation or approximated. The law of large numbers and the central limit theorem are two mathematical theorems concerning certain sampling distributions as the sample size n gets large.

Remarks

Statistics can be any numerical summary of a set of random variables; they are not limited to means, standard deviations, and proportions. Sampling distributions exist for arbitrary statistics. Only a few statistics have sampling distributions that are known. Others can be simulated through physical simulations, e.g., tossing a coin or die, or on a computer. The law of large numbers gives insight into the performance of taking averages of large samples. The central limit theorem justifies some probability calculations using the normal distribution (section 6.4). Knowing when, why, and how to apply these results is important for an applied statistician. See http://davidmlane.com/hyperstat/sampling_dist.html for additional discussion.

Outline

Section 7.1 illustrates sampling distributions of statistics in general. Section 7.2 focuses on the law of large numbers. Section 7.3 provides problems on the central limit theorem. Chapters 8 through 10 present methods for statistical inference that are based on sampling distributions.

7.1 Sampling Distributions in General

Sampling distributions with coins and dice were involved in questions 5.1 (D) and 6.1 (B). In problem 5.3 (B), if one point is added for every win and no point is added for every loss, then the sampling distribution of the number of wins in n trials under a given strategy can be simulated. Problem 7.1 (A) below uses an applet to study the properties of various statistics when sampling from assorted distributions.

Problem 7.1 (A)

Simulations of sampling distributions

The applet http://www.ruf.rice.edu/ lane/stat_sim/sampling_dist/index.html allows one to simulated sampling distributions of some statistics in various situations.

1. For normal data (mean 16, SD 5), set $n = 5$, select "Mean" for the middle graph and "Median" for the lower graph, and press "10,000 samples."

    (a) What are the averages of the sampling distributions? That is, what are the expected values of the statistics?

    (b) What are the standard deviations of the sampling distributions?

    (c) Are the sampling distributions skew or symmetric? Do they look normal?

2. Answer the questions above for the following situations.

    (a) For the skewed distribution with $n = 5$ and $n = 20$, compare the mean and median.

    (b) For the uniform distribution, compare the sd and range when $n = 5$ and $n = 20$.

    (c) For the skewed distribution, compare the sd and range when $n = 5$ and $n = 20$.

3. Create a multimodal/clustered distribution by choosing "Custom" and painting three or more clusters.

    (a) Compare the distributions of the mean and median when $n = 2$ and $n = 5$.

    (b) Compare the distributions of the sd and range when $n = 2$ and $n = 5$.

7.2 Law of Large Numbers

Problem	Description
7.2 (A)	A law of large numbers (LLN) demonstration.
7.2 (B)	Applications of the LLN: health insurance, psychic ability, and global temperatures.
7.2 (C)	Simulation of coverage of confidence intervals and the LLN.

Problem 7.2 (A)

A demonstration of the law of large numbers

The applet http://arbitrage.byu.edu/sample.html demonstrates the law of large numbers (LLN). Once you start the simulation, the program generates a sequence of random numbers from the underlying distribution.

1. Select the normal distribution for the underlying distribution.

   (a) Sample a short sequence. Is the sample mean near the population mean?

   (b) Start again, and sample a longer sequence. Is the sample mean closer to the population mean?

   (c) Start once again, and let the program run for awhile. Relate what you see to the LLN.

2. Change the distribution to uniform. Answer the questions above.

3. Change the distribution to exponential. Answer the questions above.

4. See http://statman.stat.sc.edu/~west/applets/samplemean.html. Sample several means at various sample sizes. How does the plot illustrate the law of large numbers?

Problem 7.2 (B)

Three applications of the law of large numbers

The law of large numbers (LLN), also called the law of averages, arises in several applications.

1. Health Insurance: http://ww2.northwesternmutual.com/tn/netserv–personal–page_law_large_num. If someone who is heavily insured suffers a terrible loss, an insurance company might have to pay a large settlement. How do they stay in business? Relate insurance industry profitability to the LLN.

   In INFOTRAC, search on "Law of Large Numbers" and read the article "Understanding the law of large numbers," by J. Bruce Ryan and Scott B. Clay in *Healthcare Financial Management*, October, 1995, v. 49, n. 10, pages 22-23. Describe how the law of large numbers arises in the health care industry.

2. Psychic ability: guessing a coin flip, http://statman.stat.sc.edu/~west/applets/psychictest.html.

   (a) With the probability equal to 0.5, try to guess the outcome of the coin. Do this several times. Describe how what you see relates to the law of large numbers.

   (b) Change the probability to something other than 0.5. Make a constant guess (heads or tails always) for several tosses. Describe how what you see relates to the law of large numbers.

   (c) Read about John Kerrich at http://www.anu.edu.au/nceph/surfstat/surfstat-home/3-1-1.html. How does the graph relate to the LLN?

3. Global surface temperature: See http://www.giss.nasa.gov/research/observe/surftemp. The graph presents an annual mean and a 5-year mean. Over time the global temperature is changing and slightly correlated, but the idea that averages should be expected to have less variability than individual observations applies here. Explain why the dotted line is more jagged that the solid line and relate it to the LLN.

Problem 7.2 (C)

Coverage properties of confidence intervals as an example of the law of large numbers

In chapter 8, confidence intervals are introduced. A confidence interval is a numerical summary indicating the likely location of a population mean or proportion. Confidence intervals are associated with a level of confidence, such as 95%. This means that 95% of the intervals should contain the true value of the parameter, or population value. See http://www.stat.sc.edu/~west/javahtml/ConfidenceInterval.html. Alpha is the amount below 100% coverage; i.e., 0.05 means 95% coverage.

1. Simulate 100 intervals at 95% confidence. How many cover the true value? If your answer is not exactly 95, does this mean that the law of large numbers has failed? Why or why not?

2. Simulate 1000, then 2000, then 4000 intervals at 95% confidence. What percent of your intervals cover the parameters? What do you expect in the long run?

3. Change the Alpha value and make new intervals. For you new level of confidence, what rate of coverage do you expect in the long run? What levels occurs in your simulations?

7.3   Central Limit Theorem

Problem 7.1 (A) simulated sampling distributions of various statistics under assorted distributional assumptions. In that problem, some sampling distributions were well-approximated by a normal distribution. Problems in section 6.5 illustrated the normal approximation to the binomial distribution. The binomial random variable records the number of successes in n trials, so it is equivalent to a sum of zeros (failure) and ones (success). For problems in that section, it is possible to determine for which values of p and n the normal distribution is an adequate approximation. The central limit theorem (CLT) is a general result and applies to most distributions when sums and averages of large number of independent and identically distributed units are recorded. See http://www.statisticalengineering.com/central_limit_theorem.htm, http://davidmlane.com/hyperstat/A14461.html, and http://library.thinkquest.org/10030/8stsdclt.htm for discussion of the CLT. Problems in section 7.3 are listed below.

Problem	Description
7.3 (A)	Dice demonstrations of the central limit theorem (CLT).
7.3 (B)	Examples where the CLT has some role: SAT scores and physical measurements.
7.3 (C)	An example of the limits of the central limit theorem.

Problem 7.3 (A)

Dice Demonstrations of the Central Limit Theorem

The applet http://www.amstat.org/publications/jse/v6n3/applets/CLT.html demonstrates the central limit theorem (CLT) as applied to sums of independent random variables. For parts 1, 2, and 3, briefly explain why the CLT does or does not apply.

1. Set the number of rolls equal to 10000 and the number of dice equal to 1. Roll the dice.

2. Set the number of dice equal to 2. The sum is reported. Roll the dice.

3. Set the number of dice equal to 5. Roll the dice. not.

4. Suppose you reported the average of the dice on each roll instead of the sum. Describe how the central limit theorem would apply to averages of several dice.

Problem 7.3 (B)

Two examples where the central limit theorem could have an impact

The central limit theorem (CLT) applies to situations besides dice and computer simulations. Sums and averages of observations generally have distributions that can be approximated by normal distributions. In other cases, the observations themselves are the result of the accumulation of influences of several events, processes, or forces. Examine the following situations and discuss how the CLT could apply.

1. SAT scores in the general population: Why would exam scores in the general population have an approximate normal distribution? Does the normal distribution apply to highly successful subpopulations as well? See page 2 of College Board Research Note RN-10, July 2000, "The SAT I and High School Grades: Utility in Predicting Success in College," by Wayne J. Camara and Gary Echternacht: http://www.collegeboard.com/research/abstract/1,,3869,00.html?pdf=/repository/rn10_10755.pdf.

2. Size measurements in the general population: The Center for Disease Control (CDC) clinical growth charts (see problem 6.1 (A)) present percentiles of the distribution of size measurements for boys and girls from birth to 36 months of age.

   (a) For a particular age, sex, and characteristic, based on the percentile information, graph the distribution as closely as possible. How is it similar to a normal distribution?

   (b) Why would these distributions be roughly normal? How is size determined by the "sum" of several factors?

   See also http://lib.stat.cmu.edu/DASL/Stories/ChestsizesofMilitiamen.html and problem 6.4 (A).

Problem 7.3 (C)

The limits of the central limit theorem: an example using the Poisson distribution

In INFOTRAC, search on "Poisson" and read the article "Sample sizes and the central limit theorem: the Poisson distribution as an illustration." by Ian T. Jolliffe. *The American Statistician*, August, 1995, v. 49 n. 3, page 269.

1. For what values of the Poisson mean is the normal distribution a useful approximation to the Poisson probability distribution?

2. If you add together n independent Poisson random variables each with mean λ, what is the distribution of the sum?

3. In the previous part, if $\lambda = 0.2$, how large of a sample should you collect before using the central limit theorem (CLT) to approximate the distribution of the sum?

4. If λ is 0.02 or 0.002, how large of a sample should you collect before using the CLT?

Chapter 8

Confidence Intervals

Introduction

The goal of statistical inference is to summarize what is known about the population distribution and its characteristics based on the sample values. In the repeated sampling framework, properties of sampling distributions of some statistics in some situations are known. For example, if one collects measurements on n independent and identically distributed random variables, $X_1, X_2, \ldots, X_n$, then the sampling distribution of $\bar{X}$ has the same average as the individual variables and a variance that is $(1/n)$ times the individual variance. Furthermore, if n is large, the sampling distribution is well-approximated by a normal distribution.

A confidence interval is an interval on the real number line indicating the likely location of a parameter of interest. The interval is derived through a procedure that is designed to be successful in capturing or covering the parameter value with a certain level of confidence. In the case of the sample mean, a probability statement concerning the random variable $\bar{X}$ assuming normality based on the central limit theorem can be inverted to produce an interval for μ, the population or theoretical mean. The amount of probability, such as 95%, determines the level of confidence. Similar arguments lead to an interval for p, the population proportion, that is based on the sample proportion $\hat{p}$. Linear combinations of averages and proportions from independent samples yield other formulas encountered in introductory statistics. Further discussion can be found at http://davidmlane.com/hyperstat/confidence_intervals.html.

Remarks

Several confidence interval formulas are presented in introductory statistics. The correct formula depends on how the data are recorded and how the population is sampled. In introductory statistics, inference concerning characteristics of quantitative variables generally involves means, whereas inference regarding qualitative variables is based on proportions. Constant distributions and independent samples usually are assumed. There are statistical methods for some situations involving correlated data and changing variances, but these are beyond the scope of inference as it is presented at this level. Assumptions underlying methods of inference should be acknowledged and checked to the degree possible in every application.

Context is important. The variables have units of measurement and generally are measured for some reason, to answer a question of interest. An interpretation of a confidence interval or hypothesis test should be written in a full sentence. In a report or news article, it would not be sufficient to present only numerical results. Rather, communication should be clear and in nontechnical language. A clear distinction should be made between the population under study and the sample on which data are collected. Inferential statements, however, need to recognize the repeated sampling rationale. For example, there is not a 95% chance that a particular interval covers a parameter, it either does or does not capture the population value, but the procedure has 95% coverage in repeated applications.

Outline

Section 8.1 presents problems that can be answered by forming a confidence interval for the mean of a single population. This section includes simulations aimed at explaining the meaning of "confidence level" and "coverage" of the population quantity. Section 8.2 focuses on comparing means of two populations. Section 8.3 concerns a single, unknown proportion, whereas section 8.4 contrasts two population proportions. Hypothesis testing, an alternative and complimentary approach to statistical inference, is discussed in chapter 9. Chapter 10 considers other inferential topics, including inference for the slope of a regression line. The Companion Website can be accessed at http://larsen.duxbury.com. Please enter the Serial Number from the back cover when prompted.

8.1 Confidence Intervals for a Single Mean

The data utilized in these studies are quantitative measurements on a sample of units. The methods also are appropriate for quantitative data from a matched-pairs study. The t-distribution (see http://www-gap.dcs.st-and.ac.uk/~history/Mathematicians/Gosset.html) is used for these confidence intervals when the standard deviation is estimated. Problem 7.2 (C) used simulated confidence intervals to illustrate the law of large numbers. See also http://www.math.csusb.edu/faculty/stanton/m262/confidence_means. Questions for section 8.1 are outlined below.

Problem	Description
8.1 (A)	Michelson's experiments on the speed of light and confidence intervals for a mean.
8.1 (B)	Student's t-distribution.
8.1 (C)	Confidence intervals for mean agricultural production by county.

Problem 8.1 (A)

One sample confidence intervals for the speed of light

See http://lib.stat.cmu.edu/DASL/Stories/SpeedofLight.html and the associated data file.

1. Based on reading the story, what parameter does Michelson want to estimate?

2. For each of the five trials, calculate the mean and standard deviation of the measurements. Report 95 percent confidence intervals for the speed of light.

3. How many of the intervals cover the true value? If you performed 100 experiments, all free of bias, how many typically would you expect to cover the true value of the speed of light?

4. Which intervals are sensitive to outliers? For the interval that you think would be most affected, remove an outlier and recompute the confidence interval.

What else is here? Analysis of variance (ANOVA) is used to test whether the means of two or more groups are the same. It is a generalization of a two-sample pooled t-test of section 9.2.

Problem 8.1 (B)

The t Distribution

The t-distribution arises when a sample of n independent measurements from a normal distribution with mean μ is taken and $t = (\bar{X} - \mu)/(s/\sqrt{n})$ is reported. If the population standard deviation σ were known, then $z = (\bar{X} - \mu)/(\sigma/\sqrt{n})$ would have a standard normal distribution. The t-distributions form a family of distributions indexed by the degrees of freedom, df, which equals $n - 1$ in these applications.

1. See http://www.econtools.com/jevons/java/Graphics2D/tDist.html. The t-distribution for large df is very close to the standard normal. For df 1, 3, 9, 27, and 81, what values define the central 95% of the area under the t-density? What values define this area for the standard normal? Write a sentence describing how the interval giving the central 95% area under the density changes as df increases.

2. See http://statman.stat.sc.edu/~west/applets/tdemo.html. Find the area to the right of -2, -1, 0, 1, and 2 for df 2, 10, and 50. Write a sentence describing how the area changes with df and cut-off value.

3. See http://www.stat.vt.edu/~sundar/java/applets/TNormalApplet.html. Find the interquartile range for t distribution with 2, 12, and 72 df. How does the IQR change with df?

   See also http://www.itl.nist.gov/div898/handbook/eda/section3/eda3672.htm.

Problem 8.1 (C)

Confidence intervals for mean agricultural production by county

The National Agricultural Statistics Service reports on farms in the U.S.: http://www.nass.usda.gov:81/ipedb.

1. Select "all counties in one or more states" for crops. Check boxes to select a corn for grain and the state Nebraska. Change the year beside "from" and "to" to 2001. Press the "to screen button." The number of acres planted, number harvested, and total production are three variables reported.

   (a) The average per county of the these three variables is the total, given in the last row, divided by the number of counties, which in Nebraska for this variable is 90. Compute the state averages.

   (b) Pretend now that we did not have the overall state values. Sample nine counties, record the values of the three variables, and form 95% confidence intervals. You can sample the counties by randomly generating nine integers between 1 and 90 inclusive: see the generator pages http://www.random.org/nform.html or http://www.randomizer.org/form.htm.

   (c) Write sentences explaining what these three confidence intervals mean. Do they cover the true value? *Note*: Yield, recorded in bushels per acre, for the state is not a direct average of the yields in the counties. Rather, it is a weighted average with weights proportional to harvested acres.

2. From the QuickFacts page, select "all counties in one or more states" for livestock. Select the box for Mohair (see http://www.mohairusa.com) and Texas. Set the year to 2001 and "to screen."

   (a) Two variables recorded here are number of goats and pounds of Mohair. State totals divided by the number of counties gives the average amount by county. There are 29 counties reporting for 2001. What are the average amounts? *Note*: Average fleece weight, recorded in pounds, is not a direct average of the county-level averages. It is a weighted average.

71

(b) Randomly select 3 counties and make 90% confidence intervals for the mean values. Do they cover the true values? Write sentences to present your results.

3. From the QuickFacts page, select "all counties in one or more states" for farms. Select a Iowa or Ohio, change the year to 2001, and press "to screen." Count the number of counties reporting, excluding the state total and regional totals.

    (a) The number of farms and acres in farms are reported. What are the averages per county for the state? *Note*: Acres per farm for the state is a weighted average of county averages.

    (b) Randomly sample approximately 1/10 of the counties. Form 98% confidence intervals for the state averages. Do your intervals cover the true values? Write sentences summarizing your results.

8.2   Confidence Intervals Concerning Two Means

Problems in section 8.2 concern the difference of means when data are collected from independent samples.

Problem	Description
8.2 (A)	Reading ability and directed reading activities, difference in means.
8.2 (B)	Pets and young adults: two sample confidence intervals for means.

Problem 8.2 (A)

Reading Ability and Directed Reading Activities, Difference in Means

An experiment concerning the impact of directed reading activities on reading ability of third grade students was discussed in problem 1.3 (G). See http://lib.stat.cmu.edu/DASL/Stories/ImprovingReadingAbility.html and http://lib.stat.cmu.edu/DASL/Datafiles/DRPScores.html.

1. Form a 95 percent confidence interval for the difference in the means. Write a sentence interpreting what your interval means in this example. Write a second sentence clarifying what '95 percent confidence' means.

2. Would a 99 percent confidence interval be wider or narrower? Calculate the interval and write a sentence answering the question.

3. Are the assumptions of the two-sample confidence interval for a difference of means using the t-distribution appropriate for these data? Report how you check assumptions.

Problem 8.2 (B)

Pets and young adults: two sample confidence intervals for means

In INFOTRAC, search on "pets and survey" and select "Pet attachment and generativity among young adults," by Shaela G. Marks, Jean E. Koepke, and Cheryl L. Bradley, *The Journal of Psychology*, November, 1994, v. 128, n. 6, pages 641-650. Study participants completed the Pet Attachment Survey (PAS) and Loyola Generativity Scale (LGS). Make 95% percent confidence intervals and write a sentence to answer questions below. Discuss assumptions of methods and how they would be checked using the data.

1. Compare men and women on the PAS and on the LGS.

2. Compare primary versus non primary care givers on the PAS.

3. Compare dog versus cat owners on the PAS.

4. Discuss limitations of this study. Can results can be generalized to a wider population?

8.3   Confidence Intervals for a Single Proportion

Section 8.3 presents problems using confidence intervals for a single proportion.

Problem	Description
8.3 (A)	Simulation of confidence intervals for a proportion: coverage in repeated sampling.
8.3 (B)	Flips, spins, and tips of Euro coins: what is fair?
8.3 (C)	Pet owners and their pets: confidence intervals for a proportion.

Problem 8.3 (A)

Simulation of confidence intervals for a proportion: coverage in repeated sampling
See http://www.ruf.rice.edu/ lane/stat_sim/normal_approx_conf/index.html.

1. With n=12 and p=0.5, run the simulation 3 times (30,000 samples). What percent of intervals cover p?

2. With n=12 and p=0.1, run the simulation 3 times. What percent of intervals cover p? Why is it so low?

3. With n=120 and p=0.1, run the simulation 3 times. What fraction of intervals cover the true proportion?

4. Change the confidence level to 90% or 99% and run a simulation with $n > 50$ and $0.1 < p < 0.9$. Describe the results.

Problem 8.3 (B)

Flipping the Euro: confidence intervals for proportions
See Chance News, issue 11, number 2, item 2, on the new Euro coins:
http://www.dartmouth.edu/~chance/chance_news/recent_news/chance_news_11.02.html#item2.

1. Verify the computations for the 95% confidence intervals for the proportions of heads by Buffon, Pearson, and Kerrich. Combine the samples together and make one confidence interval.

2. Make 92% and 97% confidence intervals using the data of the previous part.

3. Report 95% and 99% confidence intervals for the percent <u>tails</u> based on flips, spins and tips of the Euro.

4. Write a couple of sentences describing your results. Which coins and methods produce fair outcomes?

Problem 8.3 (C)

Pet owners and their pets: confidence intervals for a proportion
In INFOTRAC, search on "pets and survey" and select ""WOOF, WOOF" means, "I LOVE YOU"" from *American Demographics*, February 1, 2002, page 11. Assume the study collected a simple random sample.

1. Form a 95% confidence interval for the percent of pet owners who send or receive a holiday card from their cat or dog.

2. Form a 95% confidence intervals for the percent of pet owners who say "I love you" to their pet at least once per day and at least once per week.

3. Contrast the center and width of these three intervals by making a graph.

4. What sample size would be necessary in these three cases to produce a confidence interval with a half width, or margin of error, of 5%? a total width of 5%?

8.4 Confidence Intervals Concerning Two Proportions

Section 8.4 concerns confidence intervals for the difference of proportions based on two independent samples.

Problem	Description
8.4 (A)	Should women in their 40s have annual mammographies?
8.4 (B)	Pediatricians, part-time work, and confidence intervals comparing proportions.

Problem 8.4 (A)

Mammographies and confidence intervals for two proportions

Should women in the 40s have mammographies? In INFOTRAC, click on PowerTrac and enter "ke mammography and jn british medical journal" in the search window. Click on "View" and the article "Annual mammography in women in their 40s does not cut death rate," by David Spurgeon, *British Medical Journal*, September 14, 2002, v. 325, issue 7364, page 563. The article reports results in terms of counts and the ratio rates. There is ongoing debate about when and how often women should have mammographies.

1. Form a 95% confidence interval for the difference in proportions of deaths from breast cancer in the treatment and control groups. Does the interval include zero? What does this suggest about the benefit of yearly mammography beginning between 40 and 49?

2. Form and interpret a 99% confidence interval for the rate of invasive breast cancer in the two groups.

3. Form a 90% confidence interval for the difference in rates of in situ breast cancer for the two groups.

4. What assumptions underly the intervals you report? Are the assumptions satisfied in these three cases?

   For further information, see
   • "Mammography is no better than physical examination, study shows," by Deborah Josefson, *British Medical Journal*, September 30, 2000, v. 321, issue 7264, page 785,
   • (search on "ke mammography and ke benefit") "Swedish trials suggest modest benefit for screening mammograph," *Women's Health Weekly*, May 16, 2002, page 10, or "Swedish trials suggest modest benefit for screening mammography," *Cancer Weekly*, May 14, 2002, page 15, and
   • The National Cancer Institute: http://www.nci.nih.gov/cancerinfo/screening/breast.

Problem 8.4 (B)

Pediatricians, part-time work, and confidence intervals for comparing proportions

In INFOTRAC, search on "pediatricians and part-time" and select "Pediatricians working part-time: past, present, and future," by William L. Cull *et al.*, *Pediatrics*, June, 2002, v. 109, issue 6, pages 1015-1020.

1. Comparing 1993 to 2000, form a 95% confidence interval for the change in percentage of pediatricians who are female. Use the data from the two American Association of Pediatricians (AAP) surveys.

2. Comparing 1993 to 2000, how has the percent of women working part-time changed? Report and interpret a confidence interval.

3. Table 2 presents results of the 2000 3rd-year pediatrician resident study. Form confidence intervals comparing female to male residents in terms of the following reported barriers to part-time work: (a) loss of income, (b) reduction of benefits, and (c) negative influence on practice stability. Summarize the three comparisons in a couple of sentences.

Chapter 9

Hypothesis Testing

Introduction

Hypothesis testing compares hypotheses about a parameter of interest to available data and decides if one of the hypotheses can be thrown out due to its inconsistency with the data. The null hypothesis is the status quo, what already is, no change, no difference, or the current value. The alternative hypothesis is something different: a change, a new value, a special subpopulation, a new treatment or condition. If the null hypothesis were true, then the estimate should be close to the null parameter value and, in the cases considered here, the statistic should be near zero. If the null is false, then the statistic most likely will be nonzero. If the statistic is too big in absolute value, the null hypothesis is rejected. The standard for significance, the significance level or α-level, sets the limit beyond which the null hypothesis can not be entertained as plausible. See chapters 9 (http://davidmlane.com/hyperstat/logic_hypothesis.html and 10 (http://davidmlane.com/hyperstat/hypothesis_testing_se.html) of David Lane's HyperStat Online for discussion and formulas.

Remarks

Hypothesis testing can seem confusing, because the number of statements required to explain what it means is larger than for confidence intervals. Focusing on verbal explanations rather than symbols sometimes helps. The level of statistical significance determines several cutoff values: the largest P-value, the most extreme test statistic, and the most extreme estimate, such as a sample mean or proportion, for which the null hypothesis is not rejected. For a particular sample and hypothesis test, the null hypothesis is either rejected or not. This decision is either correct or incorrect. In the long run, that is, in repeated sampling, the procedure yields the correct decision $100(1 - \alpha)\%$ of the time. As always, underlying assumptions should be acknowledged and checked to the degree possible in every application.

Outline

Section 9.1 considers hypothesis tests concerning the mean of one population. Section 9.2 contains questions about the difference of means when independent samples are collected from two populations. Section 9.3 focuses on tests of a single proportion. Section 9.4 presents problems comparing proportions from two independent samples. Section 9.5 concerns the chi-square test of homogeneity in two-way tables.

9.1 Hypothesis Tests for a Single Mean

See http://www.itl.nist.gov/div898/handbook/prc/section2/prc22.htm for a discussion of hypothesis tests about a mean in engineering. Questions for section '9.1 are outlined below.

Problem	Description
9.1 (A)	Two examples: the impact of garlic bread on family interactions, the curvature of space.
9.1 (B)	Michelson's speed of light data and hypothesis tests for a mean.
9.1 (C)	Illustrations of statistical power in hypothesis testing.

Problem 9.1 (A)

One sample tests for means: the garlic bread effect and space curvature

1. See http://www.smellandtaste.org/garlic.htm.

   (a) Why is this a matched pairs study?

   (b) Interpret the results of the significance test. State hypotheses and conclusions.

   (c) Do you think that this study proves that frequent use of garlic will make the world a better place? See http://lib.stat.cmu.edu/DASL/Stories/ScentsandLearning.html a similar study.

2. In INFOTRAC, search on "curvature of space" and select "General relativity and measurement of the Lense-Thirring effect with two Earth satellites," by Ignazio Ciufolini and Erricos Pavlis, *Science*, March 27, 1998, v. 279, n. 5359, pages 2100-2103. An abstract is available through INFOTRAC. The authors used the LAGEOS (see http://www.earth.nasa.gov/history/lageos) satellite data to study the "frame dragging" (http://www.phy.duke.edu/~kolena/framedrag.html) or Lense-Thirring effect predicted by Einstein. They took measurements, computed residuals based on a model, and reported a statistic that is related to a slope. Additional work was done to account for other sources of error. The "frame dragging" theory is expressed in terms of a parameter μ.

   (a) State hypotheses concerning the authors' parameter.

   (b) What do you conclude based on the evidence?

   (c) What assumptions would be necessary to report a P-value?

   Further Study: See question one of the 2002 AP Statistics Exam. The exam is located at http://apcentral.collegeboard.com/members/article/1,3046,152-171-0-8357,00.html.

Problem 9.1 (B)

Speed of Light, revisited: hypothesis tests for a mean

Problem 8.1 (A) used Michelson's light data to make one-sample confidence intervals. Here, the data (http://lib.stat.cmu.edu/DASL/Stories/SpeedofLight.html) are used in tests of a hypothesis.

1. Based on reading the story, state a null and alternative hypothesis that is testable with Michelson's data.

2. For each of the five trials, conduct hypothesis tests of your hypotheses.

3. How many of the tests reject the null hypothesis at the 0.05 significance level? What relationship does rejection of the null have with coverage of null value in 95 percent confidence intervals?

4. Which tests are sensitive to outliers? For the trial that you think would be most affected, remove an outlier and recompute the test statistic.

Problem 9.1 (C)

Statistical power in hypothesis testing

The statistical power of a hypothesis test is the probability of rejecting the null hypothesis when it is false. Power is calculated versus a specified alternative.

1. See the applet at http://www.projects.cgu.edu/wise/power/power_applet.html concerning power when testing the mean of a normal population when the variance is known.

   (a) As the alternative mean, μ_a, is placed further away from the null mean, μ_0, what happens to power? What happens to β, the chance of a type II error, which equals 100%-Power?

   (b) Describe the impact of sample size on power and β.

2. See http://www.amstat.org/publications/jse/v6n3/applets/power.html.

   (a) As the population standard deviation, σ, decreases, what happens to power and β?

   (b) What impact does switching from a two-tailed to a one-tailed test have on power and β? Write down hypotheses, values of parameters, and levels of power. Draw pictures and label the rejection region, the region where the null hypothesis is not rejected, and the area representing power.

3. Study the material at http://davidmlane.com/hyperstat/power.html. Answer the "exercises" questions.

9.2 Hypothesis Tests Concerning Two Means

See http://www.itl.nist.gov/div898/handbook/prc/section3/prc31.htm for a discussion of hypothesis tests comparing two means in engineering. Problems included in section 9.2 are listed below.

Problem	Description
9.2 (A)	Pets and young adults: two sample hypothesis tests for means
9.2 (B)	Calcium intake and blood pressure of African-American men and other studies.
9.2 (C)	Hypothesis tests for differences in mean egg sizes of cuckoos.
9.2 (D)	Significance tests for comparing computer versus paper-and-pencil TOEFL exams.
9.2 (E)	An applet to simulate the operating characteristics of the t test.

Problem 9.2 (A)

Pets and young adults: two sample hypothesis tests for means

Problem 8.2 (B) used data from a survey of young adults to illustrate the use of two-sample confidence intervals for differences of means. Look up the article used in that problem. Answer the questions of that problem using hypothesis tests. State hypotheses in symbols and in words. Report test statistics and P-values. Discuss assumptions underlying the P-value calculations. State conclusions.

Problem 9.2 (B)

Calcium intake and blood pressure of African-American men

A double-blind study on the impact of calcium supplements on the blood pressure of African-American men is the subject of a DASL story http://lib.stat.cmu.edu/DASL/Stories/CalciumandBloodPressure.html.

1. How do you know this is a two sample design rather than a matched-pairs design?

2. What test statistic is reported on the story page? Write down the formula and say in words what the symbols mean in this example.

3. What are the pooled t-test degrees of freedom? Interpret the meaning of the P-value here.

4. What assumptions underly the P-value calculation? That is, what are the assumptions of the two-sample pooled t-test? How do the authors communicate to you their evaluation of these assumptions?

5. Look up recent articles on "blood pressure and calcium supplementation" in INFOTRAC. State hypotheses, report statistics and test statistics where possible, and describe significance test results and conclusions.

   (a) "Effect of Calcium Supplementation on Serum Cholesterol and Blood Pressure: A Randomized, Double-blind, Placebo-Controlled, Clinical Trial." (Abstract) Roberd M. Bostick. *JAMA, The Journal of the American Medical Association*, April 19, 2000, v. 283, issue 15, p. 1941.

   (b) Some articles report meta-analyses of blood pressure randomized controlled trials. Medicine in particular and science in general advances knowledge by using replication: similar studies are conducted under varying conditions. If the results are consistent, then there is greater belief in their results. How are the confidence intervals informative about hypothesis tests? See "Effect of calcium supplementation on pregnancy-induced hypertension and preeclampsia: a meta-analysis of randomized controlled trials." Heiner C. Bucher *et al. JAMA, The Journal of the American Medical Association*, April 10, 1996, v. 275, n. 14, pp. 1113-1117.

6. In INFOTRAC, search on "blood pressure and clinical trial" and select "A Clinical Trial of the Effects of Dietary Patterns on Blood Pressure" in the *Original Internist*, June, 2001, v. 8 issue 2, page 41. Describe the clinical trial, variables measured, outcomes, and significance test results.

Problem 9.2 (C)

Those tricky cuckoos: inference for differences in means

Problems 1.1 (F) and 1.2 (G) described egg-laying habits of Cuckoo birds in nests of other birds. See http://lib.stat.cmu.edu/DASL/Datafiles/cuckoodat.html for the data.

1. Choose two host birds that you are sure are raising cuckoo eggs of different average size. Conduct a two-sample pooled t-test. Report you results.

2. Choose two host birds that you are sure are raising cuckoo eggs of the same average size, or at least for which there is not a significant difference. Conduct a two-sample pooled t-test. Report you results.

3. Discuss assumptions of the two-sample t-test and whether or not the test is appropriate for these data.

Problem 9.2 (D)

Significance tests for comparing computer versus paper-and-pencil TOEFL exams.

The Test of English as a Foreign Language (TOEFL, http://toefl.org), developed by Educational Testing Service (ETS, http://ets.org), is a test of English language proficiency taken by foreign students planning to study in countries where instruction is given in English. When the test switched to a computerized format, there was a concern that familiarity with computers would impact students' English language proficiency scores. A study was conducted and is available on-line as TOEFL research report "RR-61: The Relationship Between Computer Familiarity and Performance on Computer-based TOEFL Test Tasks," by Taylor, Jamieson, Eignor, and Kirsch. March 1998: http://www.toefl.org/research/rrpts.html#rr61, PDF format, 48 pages.

1. Read the abstract (page i). Write a sentence or two describing the purpose of the study. What population is being studied? Analysis of covariance is a type of multiple regression analysis that is being used in this study to relate score on the computer-based test items to the paper-and-pencil test items while accounting for differences between students in different regions of the world. The purpose of using ANCOVA, or of making a covariance adjustment in the analysis, is to achieve a more precise estimate of the effect of computer familiarity on test results. Read pages 1-4 of the report for some background.

2. The Phase I students took the paper-and-pencil TOEFFL exam and were rated based on their responses to 11 questions in terms of computer familiarity. The Phase II students were selected from this group and completed some computerized TOEFL questions. Are the groups (Phase I and Phase II) comparable in terms of computer familiarity? See the Table 4 on page 15 of the report. Confirm the two-sample pooled t-test statistic calculations.

3. Report conclusions concerning the statistical significance of results based on the test statistics computed in the previous question. State hypotheses. What are the degrees of freedom and P-values in each case?

4. Comment on the practical significance of the differences between the groups. Paragraphs on pages 14 and 15 are relevant. Comment also on the comparison of the computer-familiar and the computer-unfamiliar students in Table 5. Are there significant differences? Are these important differences?

5. Are there differences in TOEFL scores (paper-and-pencil or computer-based) for Phase II computer-familiar and computer-unfamiliar examinees? See Table 6. Verify two of the t-statistics in the table. What degrees of freedom are used when assessing significance?

6. In Table 6, we saw that there are significant and big differences in the Phase II groups on the paper-and-pencil TOEFL tests. What implication do these differences have for comparing the performance of the Phase II groups on the computer-based questions? The ANCOVA method is used to adjust for differences between the groups in terms of paper-and-pencil test score and region of the world. Once the authors of the report make this adjustment, they conclude that of the remaining differences on the computer-based tests many are not significant and all but one interaction of factors are not of practical importance. Summary and conclusions are presented on pages 26 through 28.

Problem 9.2 (E)

The meaning of significance levels and the robustness of the t distribution.

Go to http://www.ruf.rice.edu/~lane/stat_sim/robustness/index.html. This applet simulates the performance of a two-sample t-test for the equality of means from two independent samples. Simulate 10,000 replications.

1. For the conditions below, report the estimated type I error rates and write a summary of your findings.

   (a) Population 2 standard deviation is 3, no skewness. Repeat with standard deviations 5.

   (b) Equal standard deviations and both moderately skewed. Repeat with severe skewness.

   (c) Population 2 is heavily skewed with standard deviation 3, population 1 is not skewed with standard deviation 1. Repeat with standard deviation 5 for population 2.

   (d) Increase the sample size for population 2 to 10 and repeat the previous part.

2. When the population means are not equal, the type II error rate is the fraction of the time that the null hypothesis is not rejected. Set the standard deviations to one and skewness to none.

   (a) Set the population 2 mean to 1. How does sample size affect power? Simulate for $n = 5, 10, 20$.

   (b) How does the difference between means affect power? Set the population 2 mean to 3, then 5.

   (c) How will skewness affect power? Experiment and state a conclusion.

   (d) How will standard deviation affect power? Experiment and state a conclusion.

9.3 Hypothesis Tests for a Single Proportion

Section 9.3 concerns tests of a proportion. Coverage and error rates, simulated in problems 7.2 (C), 8.3 (A), and 9.2 (E), can be made into hypothesis testing examples. The sign test in section 10.2 is another application.

Problem	Description
9.3 (A)	Flipping the Euro and other coins.
9.3 (B)	Percentages and birth characteristics.

Problem 9.3 (A)

Flipping the Euro: hypothesis tests

Chance News, issue 11, number 2, item 2, on the new Euro coins was discussed previously in problem 8.3 (B). See http://www.dartmouth.edu/~chance/chance_news/recent_news/chance_news_11.02.html#item2.

1. Test whether or not the results of Buffon, Pearson, and Kerrich differ significantly from 1/2.

2. What results would be necessary in each case to reject the null at significance level 0.05? 0.01?

3. Test whether or not the proportion of <u>tails</u> based on flips, spins and tips of the Euro is 1/2. Write a couple of sentences describing your results. Which coins and methods produce fair outcomes?

Problem 9.3 (B)

Percentages and birth characteristics

The U.S. National Center for Health Statistics (http://www.cdc.gov/nchs/births.htm) reports on all registered births. Data for 1999 and 2001 are summarized in "Births: Final Data for 1999" and "Births: Final Data for 2001" (100 page reports). We will pretend that we have random samples from 2001 and test whether or not rates have changed. Of course, the actual numbers are known and inference for 2001 is unnecessary

1. In 1999, the proportion of births by cesarean delivery was 22.0%. Suppose you have a sample of size 200 from 2001 and the sample proportion of cesarian deliveries is 24.4%. Is this a significant change? What if the sample size were 2,000? 20,000?

2. In 1998, the fraction of pre-term births (born after less than 37 weeks of gestation) was 11.8%. Suppose you have a sample of size 400 from 2001 and the fraction of pre-term births is 11.9%. Is this a significant change? What if the sample size were 40,000? 400,000?

3. In 1999, the rate of twin births was 28.9 per 1,000 births. The rate in 2001 is 30.1 per 1,000. How large of a sample size would be needed for the rate in 2001 to be judged significantly different?

4. In 2001, 12.6% of women reported smoking during pregnancy. In 2001, 12.0% women smoked. What is the power for detecting this difference at a $\alpha = 0.05$ significance level when sample size is 20,000?

5. In 2001, there were 2,057,922 male and 1,968,011 female live births. Suppose a sample of size 1,000 were taken and this rate of female live birth were observed. Would the result be significantly different from 1/2? What would be power of the test be? How about if the sample size were 10,000?

9.4 Hypothesis Tests Concerning Two Proportions

Tests concerning two proportions based on two independent samples are the focus of problems listed below.

Problem	Description
9.4 (A)	Mammographies and hypothesis test comparing two proportions.
9.4 (B)	Pediatricians, part-time work, and tests of hypotheses comparing proportions.
9.4 (C)	Computer and Internet availability and usage in the U.S.

Problem 9.4 (A)

Mammographies and hypothesis test comparing two proportions

Should women in the 40s have mammographies? Return to the article examined in problem 8.4 (A).

1. Test whether the proportions of deaths from breast cancer in the treatment and control groups differ.

2. Test for a difference in the proportions with in situ breast cancer for the two groups.

3. What assumptions underly these tests? Are the they satisfied in these two cases? Briefly explain.

Problem 9.4 (B)

Pediatricians, part-time work, and test of hypotheses

Return to the article on pediatricians and part-time work used in problem 8.4 (B).

1. Comparing 1993 to 2000, test whether the change in percentage of pediatricians who are female is significantly different from zero.

2. Comparing 1993 to 2000, how has the percent of women working part-time changed? Conduct a hypothesis test and interpret the result.

3. Table 2 presents results of the 2000 third-year pediatrician resident study. Test whether the female to male residents are different in terms of the following reported barriers to part-time work: (a) loss of income, (b) reduction of benefits, and (c) negative influence on practice stability. Summarize the three comparisons in a couple of sentences.

Problem 9.4 (C)

Hypothesis tests based on computer and Internet availability and usage in the U.S.

The Current Population Survey (http://www.bls.census.gov/cps) is a stratified multistage survey conducted by the U.S. Bureau of the Census (http://www.census.gov). Due to its design, simple averages and proportions are not reported and standard error (SE) computations are more complex than in introductory statistics. A computer ownership supplement (http://www.bls.census.gov/cps/computer/computer.htm) was conducted in September, 2001. See http://www.bls.census.gov/cps/computer/2001/ssrcacc.htm for a discussion of its SEs. Results presented in 2000 are at http://www.ntia.doc.gov/ntiahome/fttn00/contents00.html.

1. According to the report, the percent of households with Internet access rose from 26.2% in December, 1998, to 41.5% in August, 2000. Pretend that these two statistics were observed in independent simple random samples of size 300. Are the results significantly different from each other? Conduct a hypothesis test by stating hypotheses in symbols and in words, computing a test statistic and a significance level, and stating a conclusion. How would the results change if the sample sizes were 100?

2. The fraction of households with Internet access varies by type of urbanicity. In 2000, 38.9% of rural, 42.3% of urban, and 37.7% of central city households had Internet access. Pretend that these are based on simple random samples of size 425, 1350, and 775 in rural, urban, and central city areas. Test whether or not the rate is the different between pairs of areas. *Note*: Testing many pairs of groups increases the risk of erroneously finding significant results when there is no difference. See textbook commentary on "multiple testing" and "searching for significance."

3. Look at table 53 (http://www.ntia.doc.gov/ntiahome/fttn00/charts00.html#t53) on Internet use by 3- to 8-year-olds. Pretend that these data are generated from simple random samples. Test whether or not there is a difference between a pair of groups defined by time or demographic characteristics.

9.5 Hypothesis Tests Concerning Tables of Counts

Tests of the homogeneity of proportion are described at http://davidmlane.com/hyperstat/chi_square.html and http://faculty.vassar.edu/lowry/ch8pt2.html (see also ch8pt1.html). A calculator (and links to a tutorial) for chi square tests is located at http://www.georgetown.edu/faculty/ballc/webtools/web_chi.html. Questions in section 9.5 are listed below. Additional questions can be formed by arranging the data of section 9.4 into two-way tables. Data from http://lib.stat.cmu.edu/DASL/Stories/Students'Goals.html also can be tabulated.

Problem	Description
9.5 (A)	Understanding the chi-square distribution.
9.5 (B)	TOEFL examinees and computer familiarity.

Problem 9.5 (A)

The Chi square Distribution

The following problems focus on the chi square distribution.

1. See http://statman.stat.sc.edu/~west/applets/chisqdemo.html. For 1, 10, and 50 degrees of freedom,

   (a) describe the shape of the distribution.

   (b) what is the probability that a value from the distribution is greater than 5? 10? 20?

   (c) what is the chance that a randomly generated value is between 5 and 10? 10 and 20? 2 and 8?

2. See http://statman.stat.sc.edu/~west/applets/contable.html.

   (a) Compute the probabilities for the six cells under independence or constant proportions.

   (b) Compute the expected values under independence.

   (c) Compute the chi square statistic.

   (d) Double, then quadruple, the counts in the six cells. How does this change values?

Problem 9.5 (B)

The Test of English as a Foreign Language and two-way tables

Research Report 61, located at http://www.toefl.org/research/rrpts.html#rr61, was discussed in problem 9.2 (D).

1. Table 2 on page 11 (page 21 of the document) presents the number of examinees by site tested on the computerized questions. The counts of Domestic students are arranged in a three-by-two table of counts. It might be of interest to test whether or not the proportion of students in these TOEFFL groups who are computer-familiar is the same in the three cities. Given these data, you might be inclined to conduct a chi-squared test of equal proportions.

   (a) Compute the chi-square statistic, the degrees of freedom, and the P-value. What do you conclude?

   (b) Is it appropriate to conduct the test in the previous part? Read the section on "Selecting Examinees" on pages 8 and 9. Are the proportions familiar and unfamiliar with computers in the table representative of populations from these cities? Can the data available here help us answer our question of differences across cities? This problem illustrates the importance of sampling scheme for inference.

2. See Research Report 59, http://www.toefl.org/research/rrpts.html#rr59, which reports on a questionnaire administered to all TOEFL examinees in April, and to examinees in China, May, 1996. Although this is not a random sample, we will conduct hypothesis tests using these data. See tables B1-C2b.

   (a) For table C1a on examinees with low computer familiarity, are age and sex independent? Answer also for tables C2a and C3a for examinees with moderate and high familiarity, respectively.

   (b) For table C1a, are age and reason for testing independent? Answer also for tables C2a and C3a.

   (c) For table C1b, is times tested independent of TOEFL test score? Answer also for C2b and C3b.

   (d) Table B2 and B3 concern domestic and foreign test centers, respectively. Make a two-way table for one of the questions by determining to the best of your ability the number of domestic and foreign examinees answering 1, 2, 3, or 4. Test if the answer to the question and test site (domestic vs. foreign) are independent.

Chapter 10

Other Inferential Topics

Introduction

The inferential topics of introductory statistics discussed in chapters 8 and 9 constitute some of the most frequently used statistical methods. Numerous other inferential procedures are used in all areas of research. Specialized techniques are used in medicine (e.g., survival analysis), social science (e.g., logistic regression and log linear models), environmental research (e.g., spatial statistics), and economics (e.g., time series models), to name just a few. Two topics included in some versions of introductory statistics are inference in linear regression models and nonparametric tests, such as the sign test.

If a random sample is collected from a population, then inference can be performed for quantities related to correlation and regression. In a repeated sampling framework, the correlation coefficient, the regression coefficients, and predicted averages have sampling distributions. Hypothesis tests and confidence intervals can be used to perform inference for the corresponding parameters. Nonparametric statistical inference is used to avoid assumptions of standard tests. Generally nonparametric tests focus on characteristics such as medians and ranks that have universal properties. In repeated sampling inference, statistics based on nonparametric tests have sampling distributions that do not depend on a particular data distribution.

Remarks

Inferential statements can be made based on data that are sampled from a population. Ideally, the population is well-defined and the sample is randomly collected and representative. In regression, one should be aware of the definition of units being measured. Do results apply to states, counties, corporations, households, individuals or some other entities? Context and units of measurement are very important in regression, because the meaning of a slope coefficient depends on both of these factors. Nonparametric statistical tests are useful if assumptions of parametric tests are violated. Therefore, assumptions of a comparable parametric test should be evaluated. Often nonparametric methods are used in conjuction with other methods. Results for several procedures are compared. Rather than focusing on trying to make a single decision in a study, one can think of the goal of statistical inference as summarizing the data in order to learn about the population, the sampling process, and the measurement process that produced the data. When different procedures disagree, a researcher should study the data and problem and try to explain the discrepancy. Although referred to as distribution free, data structure elements such as independence of units, matched-pairs measurements, and homogeneity of populations are still important.

Outline

Section 10.1 focuses on inference for correlation and regression coefficients. Section 10.2 presents web sites related to nonparametric statistics and problems using sign tests for the median.

10.1 Correlation and Regression

Discussions of inference in correlation and regression appear at http://faculty.vassar.edu/lowry/vscor.html and http://davidmlane.com/hyperstat/prediction.html. See also the process modeling section of the NIST Engineering Statistics Handbook (http://www.itl.nist.gov/div898/handbook/pmd/pmd.htm).

Problem	Description
10.1 (A)	Inference for correlations related to brain size, acorn size, and sports statistics.
10.1 (B)	Predicting greenhouse gas emissions of cars.
10.1 (C)	The F distribution.

Problem 10.1 (A)

Inference for correlation in three problems: brain size, acorn size, and sports statistics

1. See http://lib.stat.cmu.edu/DASL/Stories/BrainSizeandIntelligence.html. For men and women,

    (a) make scatter plots and compute correlations of MRI count and both weight and height, and

    (b) report 95% confidence intervals for these correlation coefficients. Summarize your findings.

2. See http://lib.stat.cmu.edu/DASL/Stories/AcornSizeOakDistribution.html. For the two regions,

    (a) compute correlations between log acorn size and range, then

    (b) report 95% confidence intervals and write a summary.

3. Gather data on two quantitative variables for at least ten athletes in the current sports season. Record other information such league, position, and years as a professional. The more independent the observations, the better, so select one player or just a few players per team and one season per player.

    (a) Make scatter plots and compute correlations. If large differences exist between players of different leagues, positions, or experience, consider splitting the sample into homogeneous groups.

    (b) Test whether or not correlations are significantly different from zero.

Problem 10.1 (B)

Predicting greenhouse gas emissions of cars

The EPA/DOE site http://www.fueleconomy.gov (see http://www.epa.gov/epahome/cnews_0810.htm) provides consumers information on cars relating to size, miles per gallon, and emissions. Click on "Find and Compare" and select a model year, manufacturer, and model. If multiple cars appear, click on particular car.

1. Record several variables for at least twelve cars. Variables can include miles per gallon (MPG) for city and highway driving, greenhouse emissions, driver and passenger frontal and front and back seat side crash test ratings, EPA air pollution scores, and vehicle size values. What are the units of measurement?

2. Examine variables for the purpose of predicting greenhouse emissions. Report correlation coefficients and, for the best variables, scatterplots.

3. Regress greenhouse emissions on the best predictor variable. Report the regression equation.

   (a) Produce a 95% confidence interval for the slope. Interpret this interval in a sentence.

   (b) Test whether or not the slope is zero. Relate the test result to the confidence interval.

   (c) Look at a residual plot to check regression assumptions.

   See http://lib.stat.cmu.edu/DASL/Stories/carmpg.html for a related story and http://www.fueleconomy.gov/feg/FEG2000.htm for more fuel efficiency data.

Problem 10.1 (C)

The F Distribution

The F distribution is used in tests related to variability. F statistics are as ratios of variances of two groups, of model/regression sum of squares to error/residual sum of squares in linear regression, and of between/group sum of squares to within/error sum of squares in analysis of variance.

1. See: http://econtools.com/jevons/java/Graphics2D/FDist.html. How do the shape of the F distribution and the upper 5% critical value change as the denominator and numerator degrees of freedom change?

2. What are the F distribution 0.01 and 0.05 tail areas for degrees of freedom (2, 5), (2, 10), (8, 5), (8, 10), (40, 5), and (40, 10)? Draw pictures to illustrate the cutoff values. See http://www.psychstat.smsu.edu/introbook/fdist.htm, http://www.gseis.ucla.edu/courses/help/dist3.html.

3. For the pairs of degrees of freedom in the previous part, what P-value is associated with F-values of 1.25, 1.75, 2.25? See http://davidmlane.com/hyperstat/F_table.html.

10.2   Nonparametric Statistical Tests

Nonparametric tests that might be encountered in an introductory statistics course are presented at various sites on the Internet. The Hyperstat textbook (http://davidmlane.com/hyperstat/dist_free.html) includes a chapter on "distribution-free" tests. The STEPS (http://www.cas.lancs.ac.uk/glossary_v1.1/nonparam.html) glossary and chapter 9 of the ThinkQuest (http://library.thinkquest.org/10030/statcon.htm) concepts pages list several nonparametric tests. Examples can be found at http://www.wku.edu/~neal/statistics and http://www.nist.gov/speech/tests/sigtests/sigtests.htm. A sign test for a median could be used in the analysis of the DASL story on calories in food items: http://lib.stat.cmu.edu/DASL/Stories/CountingCalories.html. http://www.brettscaife.net/statistics/introstat/07nonpara describes the Mann-Whitney test for comparing two groups. http://lib.stat.cmu.edu/DASL/Stories/wasterunup.html on waste run up at Levi-Strauss plants suggests using the Mann-Whitney test or Kruskal-Wallis test for multiple groups. For other DASL examples where the Mann-Whitney test might be used, use PowerSearch (http://lib.stat.cmu.edu/cgi-bin/iform?DASL) on "Mann." Three problems on the sign test are listed below.

Problem	Description
10.2 (A)	Changes in Alaska subsistence fish catches.
10.2 (B)	Occurrence of killer tornados by state.
10.2 (C)	Ratings of retail food establishments and dining preferences of inspectors.

Problem 10.2 (A)

Alaska subsistence fishing and change over time

The Alaska Department of Fish and Game estimates the number of salmon caught by individuals for subsistence purposes. See page 30, table III-2, of the Alaska Subsistence Fisheries 1999 Annual Report (http://www.state.ak.us/local/akpages/FISH.GAME/subsist/download/asf11999.pdf) and the 1996 (http://www.state.ak.us/local/akpages/FISH.GAME/subsist/download/nwsal96.pdf) Salmon catch summary. Compute changes in catches for various species between two years for unique areas. Do not double count small areas and larger subdivisions or regions.

1. Have counts decreased? Conduct a sign test. State hypotheses, compute a test statistic, and report a conclusion.

2. Check matched-pair t-test assumptions. If they are reasonable, conduct the test and compare results.

Problem 10.2 (B)

Changes over time by state in the number of killer tornados

The National Oceanographic and Atmospheric Administration (NOAA)'s Storm Prediction Center storm reports (http://www.spc.noaa.gov/climo) include historical information on deadly tornados by state. See data from 1997 and 2002. Calculate changes in number of tornados and deaths by state for these two years. If a state is not listed, then it did not have a deadly tornado that year.

1. Have numbers of tornados decreased? Conduct a sign test assuming each state has independent observations. State hypotheses, compute a test statistic, and report a conclusion.

2. Check matched-pairs t-test assumptions. If reasonable, conduct the test and compare results.

3. Repeat the above analyses for number of deaths.

4. Are the observations independent of one another? Comment on possible dependencies.

Problem 10.2 (C)

Retail food preferences of inspectors and the sign test

In INFOTRAC, search on "sign test" and select "Grading systems for retail food facilities: preference reversals of environmental health professionals," by Owen H. Seiver and Thomas H. Hatfield *Journal of Environmental Health*, June, 2002, v. 64, issue 10, pages 8-14.

1. How are plus and minus signs determined in this example? How are the pairs formed?

2. State hypothesis related to normal versus closed restaurants at levels A, B, and C.

3. Conduct the tests of the hypotheses. Write a summary in words.

4. Write a sentence or two describing the data that would have been used in matched-pairs t-test, the hypothesis that would have been tested, and why such a test would not, presumably, be appropriate in this example.

Chapter 11

Case Studies

Introduction, Remarks, and Outline

This chapter presents four case studies. Several issues are considered and various methods are used in a single document or the study of a single phenomenon. A challenge to students and applied statisticians alike is choosing which of the available descriptive and inferential procedures and what research design to use. Usually there are many options. It is important to understand the structure and type of data involved in the problem. Are variables qualitative, i.e., categorical, or quantitative? Are the units a sample from a population? What biases are important? Are units correlated or clustered in any way? Several statistical methods can be used in every question.

Case Study	Description
11.1	Students who score at different levels in high school and on the SAT.
11.2	What affects the price of a house?
11.3	Articles related to vaccines against influenza.
11.4	Predicting the weight of horses.

Sources for additional on-line case study discussions and current examples include the *Journal of Statistics Education* (http://www.amstat.org/publications/jse, see "Teaching Bits" and "Datasets and Stories"), Chance News (http://www.dartmouth.edu/~chance/chance_news), the NIST/SEMATECH e-Handbook of Statistical Methods (http://www.itl.nist.gov/div898/handbook/casestud.htm), the UCLA Department of Statistics ((http://www.stat.ucla.edu/cases), and HyperStat Online (http://www.ruf.rice.edu/~lane/rvls, Case Studies). The Companion Website can be accessed at http://larsen.duxbury.com. Please enter the Serial Number from the back cover when prompted.

11.1 Students with Inconsistent SAT and high school grades

Which students perform well in high school, but not so well on the SAT? How are the inconsistent students different from the students who score consistently high or low? See College Board Research Note RN-15, "Students with Discrepant High School GPA and SAT I Scores," January, 2002, by Jennifer Kobrin and Glenn Milewski (http://www.collegeboard.com/research/home, search on Kobrin).

1. What are three groups being studied? That is, how are they defined and what are the sample sizes?

2. Look at table 1. The percentages in the three columns (NDS, HSD, and SATD) can be converted into tables of (approximate) counts.

   (a) Test whether gender and group membership are independent.

   (b) Are the response to the question about first spoken language and group membership independent?

   (c) Is the percent of HSD and SATD intending to major in engineering significantly different?

3. Table 2 gives summary information on grades for the groups.

   (a) Make a confidence interval for the average difference in FGPA between the NDS and HSD groups.

   (b) Conduct a hypothesis test concerning the average FGPA in the NDS and SATD groups.

4. Pick a group. Report a confidence interval for the difference between SAT-M and SAT-V.

   (a) What is the correlation between SAT-V and SAT-M? Hint: var(SAT combined) = var(SAT-V) + var(SAT-M) + 2 cov(SAT-V, SAT-M) and cov(SAT-V, SAT-M) = r sd(SAT-V) sd(SAT-M), where var is variance, cov is covariance, sd is standard deviation, and r is correlation.

   (b) What is the average and variance of the difference of SAT-V and SAT-M? Why is the variance of the difference not simply the sum of the variances? Hint: var(SAT-M - SAT-V) = var(SAT-V) + var(SAT-M) - 2 cov(SAT-V, SAT-M).

   (c) Report a 95% confidence interval for the difference in averages.

5. Consider predicting SAT-M based on SAT-V, or vice versa.

   (a) What is the regression line?

   (b) What is the mean square error (MSE) for the regression?

   (c) Report a 95 percent confidence interval for the regression slope.

6. The table 3 caption should read "Regression of FGPA on SAT and HSGPA."

   (a) Summarize why SAT combined is a better predictor than either SAT-M or SAT-V.

   (b) Summarize the performance of the regression models with SAT, HSGPA, and both predictors.

11.2   Prices of houses

Data on prices and characteristics of individual houses is available on-line in a variety of places. Additionally, state-level data are provided by the U.S. Bureau of the Census (http://www.census.gov/statab/www). Using the former data, describe how prices differ by category, e.g., one-story versus two-story house, and characteristic, e.g., amount of gross living space. The latter data assesses important factors at a macro level.

1. Many newspapers report houses and apartments (condominiums or cooperatives) for sale. On-line realtors and county assessors sometimes include detailed information on houses. Find a source of housing information and record several variables on 25-50 houses. Some variables should be quantitative and others qualitative. See http://lib.stat.cmu.edu/DASL/Datafiles/homedat.html for an example. Two realtor sites are http://realtor.com for homes for sale and http://househunt.com for homes for sale and

recently sold homes. http://www.50states.com/news provides links to newspapers. Look for a home section or real estate in classifieds. For example, http://www.chicagotribune.com/classified/realestate lists houses for sale and recent sales in the Chicago area, http://www.latimes.com/classified/realestate lists houses for sale in Los Angeles area, and http://herald.homehunter.com lists homes for sale in the Miami area. In some locations it will be necessary to limit searches to small areas and specific criteria. Some search engines list properties in price order, so sample listings to produce a broad price range.

(a) Collect data on 25-50 houses. Describe your variables.

(b) Make scatter plots of price versus quantitative variables and describe the patterns.

(c) Summarize price by levels of discrete variables. Are there significant differences?

(d) Compute regression equations and summarize your findings.

(e) Check regression assumptions with residual plots and other checks.

2. Tables 935, 937, 944, 950, and 957 in "Section 20: Construction and Housing" of the 2001 Statistical Abstract (http://www.census.gov/prod/www/statistical-abstract-02.html). reports data on the 50 states and the District of Columbia (DC). Table 957 includes home ownership in 2000 as a percent.

(a) Pick 20 or more states. Record variables for predicting percent home ownership. Totals can be turned into rates by dividing by the total number of houses (table 950).

(b) Summarize the relationship between variables.

(c) Run single and multiple regressions to predict percent home ownership.

(d) Perform diagnostics to check the appropriateness of regression assumptions.

(e) Summarize your models giving confidence intervals and sample predictions.

11.3   The effectiveness and side effects of Influenza Vaccines

Vaccines against the influenza virus are developed and given to hundreds of thousands of people in the U.S. and worldwide every year. See http://www.cdc.gov/nip/flu and http://www.fda.gov/cber/vaccines.htm for some background. Read the following articles available through INFOTRAC and answer the questions.

1. Search on "influenza and effective." Select "Zanamivir prophylaxis: An effective strategy for the prevention of influenza types A and B within households," by Arnold S. Monto *et al.*, *Journal of Infectious Diseases*, December 1, 2002, v. 186, issue 11, pages 1582-1588.

(a) Describe the study. What are the units being randomized?

(b) Test whether or not households treated with Zanamivir were less likely to have at least one case of confirmed influenza.

(c) Can a two-sample test of proportions be used to compare individuals who received Zanamivir versus placebo? Why or why not?

(d) Explain the power considerations for two-sample tests of proportions applied to households in the statistical analysis section.

(e) See table 1. Are the placebo and Zanamivir contact patient groups significantly different?

2. Search on "influenza and clinical trials." See "Clinical Signs and Symptoms Predicting Influenza Infection," by Arnold S. Monto *et al.*, *Archives of Internal Medicine*, Nov. 27, 2000, v. 160, issue 21.

   (a) Look at table 2. Compare the patients with and without laboratory-confirmed influenza.

      i. Use two-sample t-tests on age and time since onset of symptoms.

      ii. Use two-sample tests of proportions on sex, white ethnicity, and high risk status.

      iii. Use a chi squared test on time since onset classified into time periods.

   (b) Which symptoms in table 3 are significantly more likely to be seen in patients with confirmed influenza? Conduct a few tests. Comment on the issue of multiple testing in this problem.

   (c) Make a few probability trees to illustrate the probabilities in table 5. The PPV is the probability of having influenza given that the condition is present. The NPV is the chance of not having influence given that the condition is absent. See 5.2 (B) for definitions of sensitivity and specificity.

3. Search on "influenza and clinical trials." Look at "Confounding by indication in non-experimental evaluation of vaccine effectiveness: the example of prevention of influenza complications" by E. Hak *et al.*, *Journal of Epidemiology & Community Health*, December, 2002, v. 56, issue 12, pages 951-955.

   (a) Why is it difficult to conduct a randomized controlled trial of influenza vaccines?

   (b) What impact is confounding likely to have in non-experimental studies of influenza vaccines?

   (c) How can study design reduce the impact of confounding bias in these studies?

   (d) After data are collected, how can statistical methods be used to control for confounding?

11.4   Estimating the Weight of Horses

Predicting the weight of horses was considered in problem 2.3 (D). See the research articles at http://oas.okstate.edu/ojas/hapgood02.htm and http://oas.okstate.edu/ojas/hapgood01.htm.

1. What is being measured? What is the correlation of these variables with weight?

2. Express the prediction equations as a linear equations on the log scale if possible.

3. Compare R-squared values for the models. Is the graph accurate?

4. Compile data on horse weights (see problem 2.3 (D)) and fit linear and nonlinear models.

   (a) Make scatter plots on the original scale and describe the patterns.

   (b) Make scatter plots on transformed scales. Do the transformations increase linearity?

   (c) Compute correlations. Which variables are most linearly associated with weight (or log weight)?

   (d) Fit simple linear regression models and summarize results.

      i. What is the regression equation? Is the slope significantly different from zero?

      ii. What is R^2? What does it mean in this example?

      iii. What is the root Mean Square Error? Interpret its meaning by referring to a scatterplot.

      iv. Are regression assumptions reasonable for this example? Look at residual plots.

      v. Fit a regression with two predictors. Does the fit significantly improve?

Primary Web Sites

The list of web sites below is not comprehensive and does not include every Internet address referenced in this book. Rather, it is meant as an easy reference to some extensive and particularly useful sites.

- The *Internet Companion for Statistics: Guide and Activities for the Web* website can be accessed at http://larsen.duxbury.com. Please enter the Serial Number from the back cover when prompted.

- INFOTRAC: http://infotrac-college.com and http://infotrac.thomsonlearning.com.

- AP Statistics Listserv FAQ site: http://a-s.clayton.edu/apstatfaq.

  - Activities, Applets, & Datasets: http://a-s.clayton.edu/apstatfaq/activities.htm.
  - The AP Statistics exam: http://a-s.clayton.edu/apstatfaq/apexam.htm.

  and apstat-l archive site: http://mathforum.org/epigone/apstat-l.

- NCSSM Statistics Leadership Institute Notes:
  http://courses.ncssm.edu/math/Stat_Inst/Notes.htm

- U.S. Government Statistical Sites.

  - The Federal Statistics (FedStats) Homepage: http://www.fedstats.gov.
  - The U.S. Bureau of the Census: http://www.census.gov.
  - The Statistical Abstract of the United States: http://www.census.gov/statab/www.
  - The Bureau of Labor Statistics: http://www.bls.gov.
  - National Agricultural Statistical Service: http://www.usda.gov/nass.
  - The National Center for Health Statistics: http://www.cdc.gov/nchs.
  - The National Center for Education Statistics: http://nces.ed.gov.

- The American Statistical Association: http://www.amstat.org.
  Journal of Statistics Education: http://www.amstat.org/publications/jse.

- The College Board http://www.collegeboard.com.

  - College Board Research Reports: http://www.collegeboard.com/research/home.

- AP Statistics at College Board: http://apcentral.collegeboard.com.
  - Teacher's Corner, Statistics:
    http://apcentral.collegeboard.com/members/article/1,3046,151-165-0-2151,00.html.
  - AP Statistics Web Guide:
    http://apcentral.collegeboard.com/members/article/1,3046,151-165-0-21971,00.html.
  - The AP Statistics Exam Questions:
    http://apcentral.collegeboard.com/members/article/1,3046,152-171-0-8357,00.html.
  - AP Statistics Exam Questions – Free Response Analysis for 1997-2002:
    http://apcentral.collegeboard.com/article/0,3045,149-0-0-18636,00.html.

- Statistics Education and Teaching Sites

  - The Data and Story Library at Carnegie Mellon University: http://lib.stat.cmu.edu/DASL.
  - Rice Virtual Lab in Statistics: http://www.ruf.rice.edu/~lane/rvls.html.
    and Hyperstat On-line: http://davidmlane.com/hyperstat/index.html.
  - Webster West and Todd Ogden, Department of Statistics, Univeristy of South Carolina:
    http://www.stat.sc.edu/~west/applets.
    Globally Accessible Statistical Procedures: http://statman.stat.sc.edu/rsrch/gasp.
  - The Shodor Education Foundation, Project Interactivate, Probability and Statistics Activites:
    http://www.shodor.org/interactivate/activities/index.html#pro.
  - Exploring Data Web Site: http://exploringdata.cqu.edu.au.
  - VassarStats: http://faculty.vassar.edu/lowry/VassarStats.html
  - The NIST Engineering Statistics Handbook: http://www.itl.nist.gov/div898/handbook.
  - A course called Chance: http://www.dartmouth.edu/~chance
    and Chance News: http://www.dartmouth.edu/~chance/chance_news.
  - Gallery of Data Visualization: The Best and Worst of Statistical Graphics, by MichaelFriendly of
    York University: http://hotspur.psych.yorku.ca/SCS/Gallery.
  - The STEPS project: http://www.stats.gla.ac.uk/steps
    and its on-line glossary: http://www.stats.gla.ac.uk/steps/glossary.
  - Project Links: http://links.math.rpi.edu
    and its probability and statistics modules: http://links.math.rpi.edu/webhtml/PSindex.html,
    http://links.math.rpi.edu/devmodules/probstat/concepts/html.
  - SurfStat australia: http://www.anu.edu.au/nceph/surfstat/surfstat-home.
  - Charles Annis, Statistical Learning Dot Com:
    http://www.statisticalengineering.com/statistical_factoids.htm.
  - Project Wise, the Claremont Colleges' Web Interface for Statistical Education:
    http://www.projects.cgu.edu/wise.

- Probability sites

  - The Probability Web http://www.mathcs.carleton.edu/probweb/probweb.html.
  - Probability Central: http://library.thinkquest.org/11506, learning section.

- Random integer number generators, useful for sampling purposes:
  http://www.random.org/nform.html and http://www.randomizer.org/form.htm.

Subject Index

Subject	Problem(s), Sections and Case Studies

Notes

Notes

Notes

Notes

Notes

Notes